全国高等医药院校药学类实验教材

U0786509

药用植物学实验

（第二版）

主　编　路金才

副主编　贾凌云

编　者　（以姓氏笔画为序）

王　晶　孙冬雪　李　骁

贾凌云　贾景明　秦汝兰

路金才

中国医药科技出版社

内容提要

本书为全国高等医药院校药学类实验教材。全书共包括三部分内容。第一章主要介绍了基本知识与实验技术，包括显微镜的使用与保养，基本实验技术，药用植物标本的制作方法等。第二章编排了19个实验，包括显微实验，形态和分类实验等。附录包括常用试剂配制和《中国植物志》卷册索引。全书为中英文对照，以便于学生在阅读英文文献、撰写英文论文时参考。书后附有显微及药用植物的彩色照片，以便于学生在实验过程中对比参考。

本书可供高等医药院校药学类专业使用，也可作为研究生、函授生，自考生以及其他专业本科生的参考用书。

图书在版编目（CIP）数据

药用植物学实验/路金才主编 . —2 版 . —北京：中国医药科技出版社，2014. 8
全国高等医药院校药学类实验教材
ISBN 978 - 7 - 5067 - 6883 - 2

Ⅰ. ①药… Ⅱ. ①路… Ⅲ. ①药用植物学 - 实验 - 医学院校 - 教材 Ⅳ. ①Q949. 95 - 33

中国版本图书馆 CIP 数据核字（2014）第 142603 号

美术编辑 陈君杞
版式设计 郭小平

出版 中国医药科技出版社
地址 北京市海淀区文慧园北路甲 22 号
邮编 100082
电话 发行：010-62227427 邮购：010-62236938
网址 www.cmstp.com
规格 787×1092mm 1/16
印张 8 3/4
彩插 13
字数 172 千字
初版 2006 年 3 月第 1 版
版次 2014 年 8 月第 2 版
印次 2022 年 6 月第 4 次印刷
印刷 北京市密东印刷有限公司
经销 全国各地新华书店
书号 ISBN 978-7-5067-6883-2
定价 45. 00 元

全国高等医药院校药学类规划教材常务编委会

第二版前言

药用植物学是所有与植物药相关的医药学专业的基础课。实验教学在其专业学习中占有重要的地位。本实验指导是依据高等医药院校药用植物学教学大纲编写的。用于帮助学生解决实验时所遇到的问题，加强药用植物学基本技术与技能的训练，增强学生对课堂讲授内容的理解与记忆。

本书在第一版的基础上进行了修订，全书共分为两章及附录三部分内容。主要修订内容如下：根据各校实验学时调整的实际情况，部分实验进行了合并压缩，由原来的25个实验合并精减为19个实验。更换了部分插图及实验材料。新增加了显微及药用植物的彩色照片，以便于学生在实验过程中对比参考。在使用时各校可根据实验教学时数和内容进行调整，并鼓励学生根据实验大纲要求自我设计实验。

本书编写分工如下：沈阳药科大学路金才教授编写第一章，并统编全书；贾凌云副教授编写第二章实验一至实验三；孙冬雪讲师编写第二章实验四至实验六；王晶讲师编写第二章实验七至实验九；内蒙古医学院李骁副教授编写第二章实验十至实验十四；通化师范学院秦汝兰讲师编写第二章实验十五至实验十八；实验十九及附录一由沈阳药科大学贾景明教授编写。贾凌云、孙冬雪负责显微及植物外部形态附图内容。

由于我们的业务能力和水平有限，错误和疏漏之处敬请广大读者批评指正，以便改进。

编者

2014年7月

contents

目 录

第一章　基本知识与实验技术 …… (1)

Chapter 1　Elementary Knowledge and Experiment Technology …… (1)

第一节　实验室管理规则 …… (1)

Section 1　Regulations of Laboratory …… (1)

第二节　显微镜的使用与保养 …… (2)

Section 2　Operation and Maintenance of Microscope …… (8)

第三节　基本实验技术 …… (13)

Section 3　Elementary Experiment Technology …… (17)

第四节　药用植物标本的制作方法 …… (22)

Section 4　Method in Making Medicinal Plant Specimen …… (23)

第二章　实验内容 …… (26)

Chapter 2　Experiment Content …… (26)

实验一　植物细胞的构造及后含物 …… (26)

Experiment 1　Structure and Main Ergastic Substance of Plant Cells …… (28)

实验二　分生组织、保护组织与机械组织 …… (30)

Experiment 2　Merismatic Tissue, Protective Tissue and Mechanical Tissue …… (32)

实验三　输导组织、分泌组织和维管束的构造与类型 …… (35)

Experiment 3　Conducting Tissue, Secretory Tissue, Structure and Types of Vascular Bundle …… (37)

实验四　根的形态、初生构造、次生构造与异常构造 …… (40)

Experiment 4　Formation, Primary Structure, Secondary Structure and Anomalous Structure of Root …… (43)

实验五　茎的形态和变态类型，茎的初生构造，双子叶植物和裸子植物茎的次生构造 …… (48)

Experiment 5　Formation, Type and Modification of Stem; Primary Structure of Stem; Secondary Structure of Dicotyledon Stem and Gymnospermae Stem …… (51)

实验六　单子叶植物茎和根状茎的构造以及双子叶植物根状茎的异常构造 …… (56)

Experiment 6　Structure of Stem and Rhizome of Moncotyledon, Anomalous Structure of Rhizome of Dicotyledon …… (58)

实验七　叶的形态和内部构造 …… (60)

Experiment 7　Morphology and Internal Structure of Leaf …… (62)

实验八　花的形态和内部构造 …………………………………………………… (64)
Experiment 8　Morphology and Internal Structure of Flower ………………… (66)
实验九　果实、种子的形态、类型和构造 ………………………………………… (69)
Experiment 9　Formation, Type and Structure of Fruit and Seed ……………… (71)
实验十　藻、菌、地衣植物 …………………………………………………… (75)
Experiment 10　Algae, Fungi and Lichens Plant ……………………………… (76)
实验十一　苔藓植物、蕨类植物和裸子植物 ……………………………………… (78)
Experiment 11　Bryophyta, Pteridophyta and Gymnosperma ………………… (80)
实验十二　桑科、马兜铃科、蓼科、毛茛科 ……………………………………… (82)
Experiment 12　Mulberry Family (Moraceae), Birthwort Family (Aristolochiaceae), Buckwheat Family (Polygonaceae), Crowfoot Family (Ranunculaceae) ………………………………………………………………… (85)
实验十三　木兰科、芍药科、罂粟科、十字花科 ………………………………… (88)
Experiment 13　Magnolia Family (Magnoliaceae), Peony Family (Paeoniaceae), Poppy Family (Papaveraceae), Mustard Family (Cruciferae) ………………………………………………………………… (90)
实验十四　蔷薇科、豆科、五加科 ……………………………………………… (93)
Experiment 14　Rose Family (Rosaceae), Bean Family (Leguminosae) and Ginseng Family (Araliaceae) …………………………………………… (96)
实验十五　伞形科、木犀科、唇形科、茄科 ……………………………………… (99)
Experiment 15　Carrot Family (Umbelliferae), Olive Family (Oleaceae), Mint Family (Labiatae), Nightshade Family (Solanaceae) …………… (102)
实验十六　玄参科、桔梗科、菊科 ……………………………………………… (105)
Experiment 16　Figwort Family (Scrophulariaceae), Bellflower Family (Campanulaceae), Compositae Family (Compositae) …………………………… (107)
实验十七　禾本科、天南星科、百合科 ………………………………………… (110)
Experiment 17　Grass family (Gramineae), Aroid family (Araceae), Lily family (Liliaceae) ……………………………………………………… (112)
实验十八　薯蓣科、鸢尾科、姜科、兰科 ……………………………………… (115)
Experiment 18　Yam Family (Dioscoreaceae), Iris Family (Iridaceae), Ginger Family (Zingiberaceae), Orchid Family (Orchidaceae) ……… (117)
实验十九　烟草叶片愈伤诱导和植株再生 ……………………………………… (119)
Experiment 19　Callus Induction and Plant Regeneration From Leaf Explants of Tobacco ………………………………………………………… (121)
附录 ……………………………………………………………………… (123)
Appendix ………………………………………………………………… (123)
一、常用试剂的配制 ………………………………………………………… (123)
二、《中国植物志》卷册索引 ………………………………………………… (126)

第一章　基本知识与实验技术

Chapter 1　Elementary Knowledge and Experiment Technology

第一节　实验室管理规则

（1）遵守实验课时间，不得无故缺席、迟到或早退。如确实中途有事需要离开实验室，必须征得教师同意。

（2）每次实验内容事先应充分预习，并备好应携带之物品及有关参考资料（包括实验指导及教材）。

（3）爱护实验室内一切公共财物。使用显微镜必须严格遵守操作规程，每次使用完毕，要擦拭干净，妥善放置，如有损坏，应及时主动报告，以明确责任。

（4）爱护一切标本、样品（显微镜切片等），观察后必须归还，不得私自携带出室外（如有破损，按价赔偿）。

（5）实验室按学号安排座位，固定不变，若需调换座位须报告老师，实验中各组使用自己的实验用具、药品、标本，避免各组间相互混用。

（6）具腐蚀性液体必须倒在废液缸内，不得倒入下水道中，废物应放入垃圾桶中。

（7）除特殊情况经教师许可外，实验报告应在实验结束后交给实验指导教师，经检查后方可离开。

（8）做完实验后，个人应将自己使用的用具擦洗干净，收入盒中，妥为保管，并将自己的桌面打扫干净。

（9）实验结束后，公共实验用具应归还原处，实验桌面应收拾干净，洗净仪器、用具，关好水电，方可离开实验室。

（10）值日同学负责清扫卫生，检查煤气、水、电及门窗，保证安全整洁。

Section 1　Regulations of Laboratory

1. Be punctual to the laboratory, and do not to be absent or be late or leave early. If you have to go out for something and can not continue the experiment, get the teacher's permission first.

2. Prepare every experiments readily, and do remember to bring the materials and references (including the experimentalguide and the textbook) with you.

3. Take care of all the public properties in the laboratory. Comply with the operating rules strictly when using the microscope. After each use, make it clean, displace it correctly. If there's any damage, note the teacher in time so as to identify the responsibility.

4. Take care of all the specimens and samples (such as microscopeslice). Return it after observation and do not take them out of laboratory without permission. (If there's any damage, pay for it by the price)

5. We arrange the seats by the school number and for permanently. If you want a change, you have to ask the teacher. Use your own experiment tools, reagents and specimens, and do not misuse them with others.

6. Pour the corroded liquid into the waste jar, and do not pour it into drains. Throw the waste materials into the waste basket.

7. Unless there's special occasion and with the teacher's permission, the report of each experiments should be handed in to the guide teacher on spot and you may leave after it checked.

8. Once the experiment is over, clean up all your used tools, put away them into boxes, deposit them with care and make your table tidy and clean.

9. Return the public tools to the same place as they used to be after experiment. Make sure that the table, apparatus and tools are clean and the water and electricity is powered off before you leave.

10. The students on duty should clean up the laboratory and check the gas, water, electricity is powered off, the windows and doors are shut and the whole laboratory is clean and in order.

第二节　显微镜的使用与保养

一、显微镜的类型

显微镜是研究植物细胞结构、组织特征和器官构造最常用的重要仪器。显微镜的类型很多，根据所用光源的不同可分为光学显微镜和电子显微镜两大类。

1. 光学显微镜　由一组光学放大系统和机械支持及调节系统组成。是以可见光（紫外线显微镜以紫外光）作光源，用玻璃制作透镜的显微镜。可分为单式显微镜与复式显微镜两类。单式显微镜结构简单，常用的如放大镜，由一个透镜组成，放大倍数在10倍以下。解剖显微镜为构造稍复杂的单式显微镜，是由几个透镜组成，放大倍数在200倍以下。复式显微镜结构较复杂，至少有两组以上透镜组成，是植物形态解剖实验最常用的显微镜。其有效放大倍数可达1250倍，最高分辨率为0.2μm（1μm＝1/1000mm）。

光学显微镜又可分为普通光学显微镜、紫外线显微镜、荧光显微镜、暗视野显微镜、相差显微镜和微分干涉差显微镜等类型。

2. 电子显微镜　是使用电子束作光源的一类显微镜。电子显微镜以特殊的电极和磁极作为透镜代替玻璃透镜，能分辨相距2Å左右的物体（1Å = 1/10000μm），放大倍数可达80 ~ 120万倍，其分辨率比光学显微镜大1000倍，是观察植物细胞超微结构的重要精密仪器。电子显微镜可分为透射电子显微镜、扫描电子显微镜和扫描隧道显微镜等。

二、复式显微镜的构造

常用的复式显微镜有单筒镜和双筒镜两种（图1 - 1，图1 - 2），这两种显微镜的基本构造都包括保证成像的光学系统和用以装置光学系统的机械部分。现以简单的单筒复式显微镜为例介绍其构造。

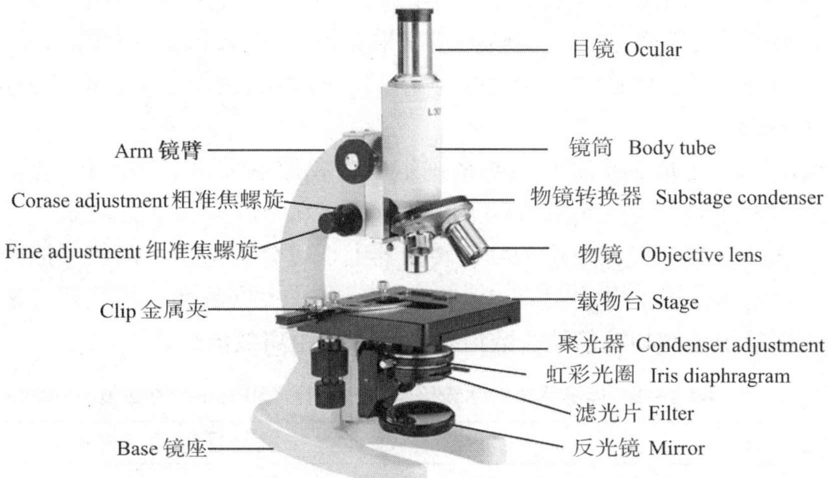

图1 - 1　单筒复式显微镜构造

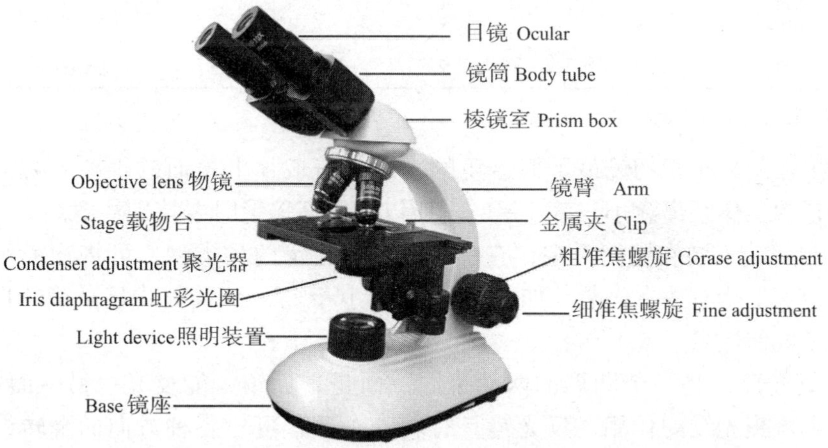

图1 - 2　双筒复式显微镜构造

1. 机械部分

（1）镜座　显微镜的底座，用以支持整个镜体，使显微镜放置稳固。

（2）镜柱　镜座上面直立的短柱，支持镜体上部的各部分。

（3）镜臂　早期简单的显微镜的镜臂弯曲如臂，下连镜柱，上连镜筒，为取放镜体时手握的部分。现在镜柱与镜臂大多为一体。

（4）镜筒　显微镜上部圆形中空的长筒，其上端置目镜，下端与物镜转换器相连，并使目镜和物镜的配合保持一定距离。镜筒能保护成像的光路和亮度。

（5）物镜转换器　为接于镜筒下端的圆盘，可自由转动。盘上有3~4个安装物镜的螺旋孔。当旋转物镜转换器时，物镜即可固定在使用的位置上，保证物镜与目镜的光线合轴。

（6）载物台（镜台）为放置玻片标本的平台，中央有一通光孔。上面有玻片卡尺，一方面可固定玻片标本，同时可以向前后左右各方向移动。

（7）调焦装置　用以调节物镜和标本之间的距离，以便得到清晰的物像。在镜柱两侧有粗、细调节螺旋各一对，旋转时可使载物台上升或下降，大的一对为粗调焦螺旋，旋转一圈可使载物台移动2mm左右。小的一对为细调焦螺旋，旋转一周可使载物台移动约0.1mm。

2. 光学部分　由成像系统和照明系统组成。成像系统包括物镜和目镜，照明系统包括反光镜和聚光器。而电光源的显微镜自身具有发光灯不需要反光镜。

（1）物镜　安装在镜筒下端的物镜转换器上，可分为低倍、高倍和油浸物镜3种。物镜可将被检物体作第1次放大，一般其上均刻有放大倍数和数值孔径（N.A），即镜口率，常用标准物镜的物镜倍数、数值孔径和工作距离数值见表1-1。

表1-1　标准物镜的物镜倍数、数值孔径和工作距离数值

物镜倍数	数值孔径（N.A）	工作距离（mm）
10×	0.28	6.5
20×	0.50	2.0
40×	0.65	0.60
80×	1.25	0.20
100×	1.25	0.198

注：100×物镜为油浸物镜。

所谓工作距离是指物镜最下面透镜的表面与盖玻片上表面的距离。物镜的放大倍数愈高，它的工作距离愈小（表1-1）。所以使用高倍镜时要特别注意。

（2）目镜　安装在镜筒上端，有单目镜和双目镜两种类型，可将物镜所成的像进一步放大，其上刻有放大倍数，如5×，10×，16×等。双筒复式显微镜的目镜可以调焦，提高了成像质量。

（3）反光镜　是一个圆形的两面镜。一面是平面镜，能反光；另一面是凹面镜，兼有反光和聚集光线的作用。反光镜具有转动关节，可作各种方向的翻转，将光线反射在聚光器上。

（4）聚光器　装于载物台下，由聚光镜和虹彩光圈等组成，它可将平行的光线汇集成束，集中于一点以增强被检物体的照明。

（5）虹彩光圈　装在聚光器内，扳动操作柄，可调节光圈大小，控制透光量。

三、显微镜的使用方法

1. 取镜和放置　每次实验时，要按固定编号从镜柜里取出显微镜。取镜时应右手握住镜臂，左手平托镜座，保持镜体直立，严禁用单手提着显微镜走动，以防止螺旋脱扣，部件散落。放置桌上时，一般应放在座位的左侧，距桌边约 10～15cm 处，以便于观察和防止掉落。

2. 对光　一般常用日光灯作光源，也可用由窗口进入的散射光，避免用直射阳光。对光时先把低倍物镜转到中央，对准载物台上的通光孔，然后用左眼或双眼从目镜向下注视，同时，转动反光镜，使镜面向着光源，光弱时可用凹面镜。当在镜筒内见到一个圆形而明亮的视野时，再利用虹彩光圈调节光的强度，使视野内的光线均匀而明亮。对于电光源的显微镜，直接打开光源开关，然后调节光的强度即可。

3. 低倍镜的使用　低倍镜的视野相对较大，容易发现目标和确定要观察的部位。所以观察任何标本，都必须先用低倍镜。其基本操作过程如下。

（1）放置切片　显微镜对光调好后，降低载物台，把玻片标本放在载物台中央，使材料正对通光孔，然后用移动器或卡尺固定在载玻片的两端。

（2）调整焦点　两眼从侧面注视物镜，并慢慢按顺时针方向转动粗调焦螺旋，使载物台徐徐上升至物镜离玻片约 5mm 处。用左眼或双目注视目镜筒内，同时按逆时针方向转动粗调焦螺旋使载物台下降，直到看见清晰的物像为止（注意不可在调焦时边观察边上升载物台，否则会使物镜和玻片触碰，压碎玻片，损伤物镜）。如一次看不到物像，应重新检查材料是否放在光轴上，重新移正材料，再重复上述操作过程直至物像出现和清晰为止。

为了使物像更加清晰，此时可轻微转动细调焦螺旋使物像最清晰。当细调焦螺旋向上或向下转不动时，即表明已达极限，且勿再硬拧，而应重新调节粗调焦螺旋，拉开物镜与标本间的距离，再反拧细调焦螺旋，约 10 圈左右（一般可动范围为 20 圈）。有的显微镜可把微调基线拧到指示微调范围的二条白线之间，再重新调整焦点至物像清晰为止。

（3）低倍镜的观察　焦距调好后，可根据需要，前后左右移动玻片使要观察的部分在最佳位置上。找到物像后，还可根据材料的厚薄、颜色、成像反差强弱是否合适等再调节，如视野太亮，可缩小虹彩光圈，反之则可开大光圈。

4. 高倍镜的使用

（1）选好目标　因高倍物镜只能将低倍镜视野中心的一部分加以放大，故在使用高倍镜前应在低倍镜中选好目标并将其移至视野的中央，转动物镜转换器，把低倍物镜移开，换上高倍物镜，使之与镜筒成一直线（因高倍镜工作距离很短，操作要小心，防止镜头碰击玻片）。

（2）调整焦点　在正常情况下，当高倍物镜转正之后，在视野中即可见模糊物像，只要稍调动细调焦螺旋，即可见到最清晰的物像。

初用一台显微镜时，要注意它的高、低倍物镜是否能如上述情况很好配合，如果高倍镜离盖玻片较远看不到物像时，则需重新调整焦点；此时应从侧面注视物镜，并小心转动粗调焦螺旋使载物台慢慢上升到高倍物镜头几乎要与切片接触时为止（小心勿压碎玻片标本和损毁镜头），然后再由目镜观察，同时转动粗调焦螺旋，稍微降低载物台至见到物像后，换调细调焦螺旋，使物像更加清晰为止。

（3）调节亮度　在换用高倍镜观察时，视野变小变暗，所以要重新调节视野的亮度，此时应该放大虹彩光圈或用凹面镜以增加进光量。

5. 油镜的使用　在使用油镜之前，也要先用低倍镜找到被检部分，再变换成高倍镜调整焦点，并将被检部分移到视野中心，然后再换用油镜。

使用油镜时，要先在盖玻片上滴加一滴香柏油，才能使用。用油镜观察标本时，绝对不许使用粗调焦螺旋，只能用细调焦螺旋调节焦点。如盖玻片过厚，必须换成薄片方可聚焦，否则会压碎玻片，损伤镜头。

油镜使用后，应立即以拭镜纸蘸少许清洁剂［乙醚和无水乙醇（7∶3）的混合液］擦去镜头上的油迹。

6. 显微镜使用后的整理　观察结束，应先降低载物台，取下玻片，转动物镜转换器，使物镜镜头与通光孔错开，再升高载物台，并将反光镜还原成与桌面垂直，擦净镜体，罩上防尘罩。仍用右手握住镜臂，左手平托镜体，按号放回镜柜中。

四、指针的安装与测微尺的使用

1. 安装指针的简易方法　如果新购置的显微镜没有指针，可以自行安装，具体方法是先将目镜的上盖旋下，剪取5~10mm一段头发，用镊子夹住，在另一头蘸上加拿大树胶，将其粘贴在目境内壁的金属铁圈上，并使指针的尖端位于视野的中央，稍干后，旋紧上盖即可使用。

2. 测微尺的使用

（1）台式测微尺　是一种特制的载玻片，中央有一个圆圈，圆圈内有一个具刻度的标尺（图1-3），全长为1mm，共分为100小格，每小格长0.01mm，即10μm。

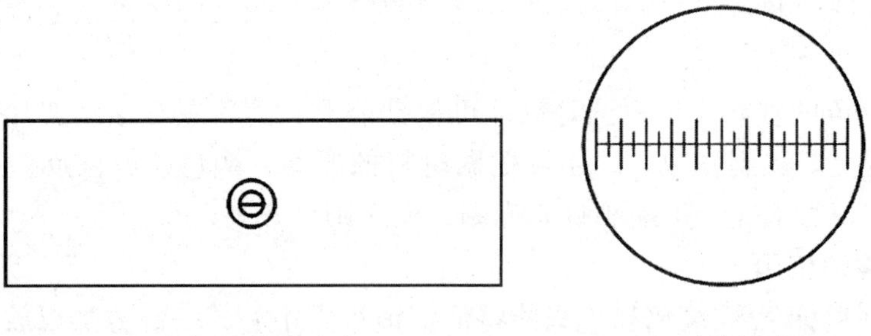

具标尺的载玻片　　标尺的放大

图1-3　台式测微尺

（2）目镜测微尺　是放在目境内的一种标尺，为一块圆形的玻璃片，直径20~21mm，上面刻有不同形式的标尺。有直线式和网格式两种（图1-4）。测量长度一般用直线式，共长10mm，分成10大格，每大格又分成10小格，共100小格。网格式测微尺用于计算数目和测量面积。

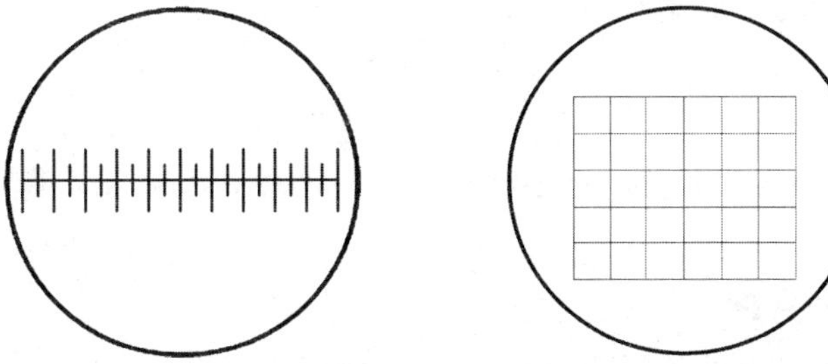

直线式目镜测微尺　　　　网格式目镜测微尺

图 1－4　目镜测微尺

(3) 细胞及细胞内含物等的测量　先将目镜测微尺装入目境内的铁圈上，用镜台测微尺标化。标化时，转动目镜，移动镜台测微尺，使两种量尺的刻度平行，并使它们的一端重合（图 1－5），再找出另一端的重合刻度，分别记录目镜测微尺和镜台测微尺重合范围内的刻度，计算出目镜测微尺每小格在该物镜条件下所相当的大小（μm）。如用 5 × 目镜和 40 × 物镜，测得目镜测微尺 100 格，等于镜台测微尺的 50 格，即目镜测微尺在这一目镜和物镜组合中每格实际长度为 5μm。测量时，以目镜测微尺测量被检物的小格数，乘以每小格的大小（μm）即得。如果用不同倍数的目镜，必须重新标化和计算。

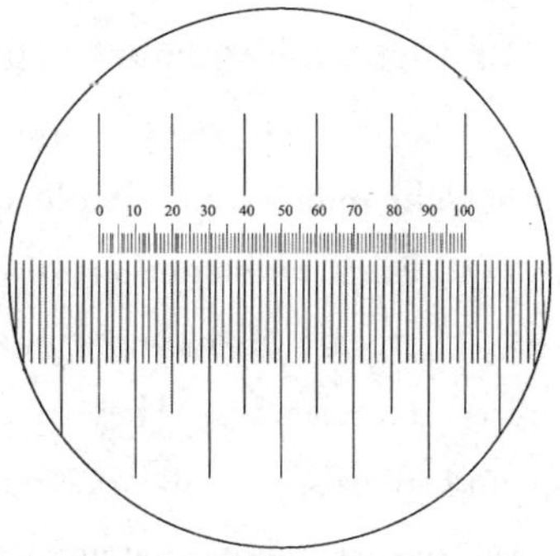

图 1－5 测定目镜测微尺每格的实际长度

（上方的标尺为目镜测微尺，下方为镜台测微尺，目镜测微尺的 100 格与镜台测微尺的 50 格重合）

$$\text{目镜测微尺每格的值} = \frac{\text{两重合线间镜台测微尺的格数} \times 10\mu m}{\text{两重合线间目镜测微尺的格数}}$$

五、显微镜的使用和保管注意事项

(1) 应随时保持清洁，机械部分可用软毛巾擦拭。光学部分的灰尘必须用镜头毛刷拂去，或用吹风球吹去，再用拭镜纸轻擦，切忌用手指或其他粗糙物如纱布等擦拭，以免损毁镜面。

(2) 在桌面上移动显微镜时，应轻拿轻放，切忌拖拉震动显微镜。

(3) 用显微镜观察时，必须用双眼，切勿紧闭一眼。要反复训练用左眼窥镜，右眼作图。

(4) 所观察标本必须加盖盖玻片，制作带水或试剂的玻片标本，必须两面擦干再放载物台上观察。

(5) 如遇机件失灵，使用困难时，不可强行转动，更不能任意拆修，应立即报告指导教师解决，以免造成损坏。

(6) 实验结束后要仔细擦拭显微镜，要有责任心。

（7）显微镜不用时应加塑料罩，及时放回镜柜内。镜柜内应放一小袋硅胶干燥剂防潮。

Section 2 Operation and Maintenance of Microscope

1. Type of microscope

The microscope is one of the important pieces of equipments that are used in studying the cell structure, tissue features and organ constitutions of plants. There are some different kinds of microscopes in use, according to the difference of optical resource, we can classify them as two kinds: light microscope and electron microscope.

1.1 Light microscope This kind of microscope is made of lenses, which use visible light (ultraviolet ray microscope use ultraviolet ray) as the optical resource. There are two types of light microscope simple ones and compound ones. The simple microscopes are made up easily. Take the magnifier for example, it is made of a single lens, with a magnification under 10. There are also simple microscopes with more complicated structure as we call them anatomized microscopes. They are made of several lenses, with a magnification under 200. The compound microscopes have complicated features, with at least two groups of lenses and are often used in the plant anatomizing experiments. Their effective magnification reaches up to 1250, with the resolution 0.2μm (1μm = 1/1000mm).

Except for these common microscopes used in the primary lab, there are also some kinds used in research work such as Ultraviolet Ray Microscopes, Fluorescence Microscopes, Dark - field Microscopes, Phase - contrast Microscopes and Differential Interference Contrast (DIC) microscopes.

1.2 Electron microscope

This type of microscope uses a beam of electrons rather than visible or ultraviolet light and can differentiate any substance upon a distance 2Å (1Å = 1/10000μm). The magnification can reach up to 0.8 - 1.2 million, 1000 times larger than a light microscope. It is an important exact apparatus to study the micro - world. Electron microscopes includes Transmission Electron Microscopes, Scanning Electron Microscopes and Scanning Tunneling Microscopes (STM).

2. The structure of compound microscope

The compound microscopes usually have one drawtube or two drawtubes (Figure 1 - 1 and 1 - 2). Both of the two kinds above have an optics system which assure us to image and a mechanical system which the optics system fitting in. Now we take a simple and one drawtube microscope for example to introduce its components.

2.1 Mechanical system

2.1.1 Base The foundation of the microscope uphold the whole apparatus and makes it steady.

2. 1. 2 Column　A erect short column on the base, upholds the department above it.

2. 1. 3 Arm　It incarnates as an arm, connect the base and the column. We hold it when taking the microscope.

2. 1. 4 Body tube　This is a long tube to separate the objective lens and the ocular. It can protect the light way and brightness of the image.

2. 1. 5 Substage condenser　A disk under the column, which can be turned easily . There are 3 – 4 screw poles on it. When it is turned, the objective lens can be fixed at a certain place, and make the light of the objective lens and the ocular lens on a straight line.

2. 1. 6 Stage　On which the specimen is placed. There is a core in the central, through which the light pass away. It has a turner on it, which can fix the slide and move on each position.

2. 1. 7 Adjustment　Raise and lower the objective lens so as to bring the specimen into focus There are a couple of coarse adjustments and another couple of fine adjustments. They can move the stage up and down when turned. The bigger ones are coarse adjustments, which can move the stage 2mm when they are turned a circle. The fine adjustments are much smaller and can move the stage 0. 1mm.

2. 2 Optics system

This is made up of the imaging system and illuminating system. The imaging system contains an objective and an ocular. The illuminating system contains a mirror and a condenser adjustment. There is not a mirror in the microscope use electric light.

2. 2. 1 Objective lenses　It is settled on the substage condenser and can be classified as lower – power, high – power, and oil – immersion. It magnifies the specimen for the first step. Commonly, it has the magnitude times work distance and the number of aperture（N. A）. There are some data of standard objective lenses in Table1 – 1 .

Table 1 – 1 Some data of standard objective lenses

Multitude of objective lens	The Number of aperture（N. A）	Work distance（mm）
10 ×	0. 28	6. 5
20 ×	0. 50	2. 0
40 ×	0. 65	0. 60
80 ×	1. 25	0. 20
100 ×	1. 25	0. 198

100 × objective lens: oil – immersion objective lens.

The work distance is defined as the distance between the surface of the down most lens and the surface of the slide on the specimen. The greater the magnitude time, the smaller the work distance will be（Table 1 – 1）. So it needs more care when we use high – power objective lens.

2. 2. 2 Ocular　It is settled on the top of the column and can magnify the image from the objective. It has magnitude times on it such as 5 × , 10 × and 16 ×. Ocular have one drawtube

and two drawtube two types. The two drawtubes ocular can be adjusted to improve the quality of image.

2. 2. 3 Mirror It is a two – sided rounded mirror. The one side is a plane mirror, and the other side is a concave mirror, which both can reflect and collect the light. The mirror has a turner which can be turned to all sides and reflects the light right on the condenser adjustment.

2. 2. 4 Condenser adjustment This is settled under the stage and is made up of a condenser mirror, an iris diaphragm and so on. It can collect the parallel light into a beam to strengthen the illumination on the specimen.

2. 2. 5 Iris diaphragm It is settled in the condenser adjustment. We can adjust the side of the diaphragm by drawing the operation bar so as to control the light overpasses.

3. The use of microscope

3. 1 Fetch and display Each time, please get out of the microscope from the cupboard with the right number. You should hold the arm with your right hand and support the base with your left hand to make the whole body steady. You mustn't lift the microscope with single hand, or the condenser may break off. The microscope should be put on the left side of your table with a distance from the table edge 10 – 15cm in order to observe and prevent it from falling down.

3. 2 Adjustment of the light Usually, we use the light shining from the window or the light from daylight lamp, not direct shining (point black) sunlight. First turn the low power lens to the central right onto the pore and then observe the specimen with your left eye or both. Meanwhile, turn the mirror to make the surface face to the light. You may use the concave mirror when the light is weak. Once you see the bright round vision inside, turn the iris diaphragm for better light power to make the vision clear.

3. 3 The use of low – power objective lens Whenever you observe a specimen, you should start with the low – power object lens. The low – power object lens can give a wider vision field, and you can find the destination and assure the exact part of the objective you want easily.

3. 3. 1 Lay the slice up Lower the stage, set the specimen right on the central of the stage and make the point you want right to the pore, and then fixed the two sides of the slice with the remover.

3. 3. 2 Adjust the focus Look at the objective lens with your eyes from the profile, slowly turn the coarse adjustments close widely to raise the stage until the distance between the stage and the slide is about 5mm. Observe the vision with your left eye or both of your eyes and turn the coarse adjustments counterclockwise until you see the image precisely. (Don't lift up the stage while you observe, or you may smash the objective lens and the slide, break the lens or the glasses.) If you can't see the image, check whether your material is right on the light axes, remove it and do the above again till you see what you want clearly.

To make the image more clear, turn the fine adjustment slightly. When the fine adjustment can't move (up or down) any more, it has reached its limit. Don't move it any more, just readjust your coarse again, enlarge the distance between the objective and the specimen, and then turn your fine adjustment towards the other side about 10 revolution (At large, it can be turned 20 revolution to the limit). Some microscopes can adjust the baseline within the two limit lines at both sides, and then adjust the focus to get a clear vision.

3.3.3 Observe with the low – power objective lens After the focus is adjusted, you can move the slide to the best place as you want to. You could also adjust it according to the thickness and colour of your material and the contrast intension of the image after you find the image. For instance, if the vision is too light, turn down the iris diaphragm. Turn it up when at reverse.

3.4 Use of high – power objective lens

3.4.1 Mark your destination Before using the high – power objective lens, you should make your destination right in the central of the field of vision with the low – power objective lens because the former can magnify only a part by the central vision of the latter. Then turn the low – power lens away, with the high – power lens instead. Make sure that it holds a line with the body tube (The work distance is short in high – power, so take care with your operation, preventing from smashing the lens and the slide).

3.4.2 Adjust the focus As the high – power turns right, we can see the vision vaguely. After adjust the fine adjustment a little, we can see it clearly.

When you start with amicroscope, make sure that it works well as the situation above. If you can't see the image with your high – power objective at the original place, readjust your focus. Observe the objective from profile, turn the coarse adjustment up slightly till it is going to touch the slice (Be sure not to break either the slice up or the lens), and then observe it with the oracle. Turn the coarse adjustment, lower the stage till you see the image, and turn the fine adjustment again to make the image more clearly.

3.4.3 Adjust the luminance When you turn into the high – power, the field of vision will be smaller and darker. Readjust the luminance of the vision by magnifying the iris diaphragm or using the concave mirror.

3.5 Use of the oil – immersion Before using the oil – immersion, find the part you want with the low – power objective, adjust the focus with the high – power one, take the part to the central field of vision, and then turn it to the oil – immersion.

Drip a drop of cedar oil on the top glass before using the oil immersion. You mustn't use the coarse adjustment when observe with oil – immersion, but only the fine adjustment. If the top glass is too thick, alternate a thin one before adjusting the focus, or it will smash the slide or the lens.

After using the oil – immersion, tip it with a few drops of cleanser immediately (a mixture of ether and alcohol (7∶3)) to wipe off the oil spot.

3.6 Arrangement after using After using the microscope, lower the stage, get the slice up off, turn the condenser not to make the objective lens and the pore on line, lift the stage again, make the mirror upright with the table, make the whole body clean and cover the dust-proof cover. Get microscope back to the cupboard in the same way you brought it out.

4. The settlement of pointer and the use of the micrometer

4.1 The settlement of the pointer If there is not any pointer in yourmicroscope, you can settle one by yourself. Turn off the top cover of the ocular, cut down a tiny bit (5 - 10mm) of your hair, clip it with a clipper, dip it into the Canadian gum at one end, paste it onto the metal coil fixed on the inner wall of the ocular (make sure that the point at the central of the field of vision) and tighten the top cover on after the gum dried. It is ready to use now.

4.2 Use of the micrometer

4.2.1 Stage micrometer A special slide, with a graduated meter in the central, with the length 1mm, 100 division with each 0.01mm (10μm) (Figure 1 - 3).

4.2.2 Ocular micrometer A meter in the ocular, It is a disc slide, with a diameter of 20 - 21mm. There are two kinds of staff gauges, lineal ones and gridding ones (Figure 1 - 4). The former are used to measure the length, with a length 10mm, 10 divided, with each portion 10 divided again, which means 100 share of the whole. The latter are used to count the number and the area.

4.2.3 The measure of the cell and the cell inclusion Settle the ocular micrometer onto the metal coil, and standardize it with the stage micrometer. During the standardization, turn the ocular, move the stage micrometer, make the two graduation parallel with an end accor - dant to each other, find the other end's superposition and mark the graduations that the two accor dant with one another during the scope. Calculate the length of each share on the ocular micrometer amount to the stage micrometer (μm). With the 5 ×, 10 × objective, 100 shares on the ocular micrometer amount to 50 shares on the stage micrometer which is to say, the length of each share on the ocular micrometer is 5μm. When measuring something, we use the ocular micrometer to get the share number, and multiple the length of each share to get the result. If we use different kinds of oculars, we must standardize and calculate them again.

5. Attention to use and storage of microscope

5.1 Keep the microscope tidy all the times, wipe the mechanical department clean with a soft washcloth, and clean the ocular department with a lens - used brush or blow the dust away before you wipe it with a lens - cleaned paper. Be sure not to wipe it with your hands or some other coarse things like cotton or you may scrape the lens.

5.2 When to move a microscope on the table, you should take it slightly, be sure not to pull it on the table, which can shake it roughly.

5.3 Use both of your eyes when observing, and do not close either of your eyes. Always practice to use your left eye for observing, the other for drawing.

5.4 You must put a top glass on the specimen before observing the stage. If you make the slice up with water or other solutions, be sure to wipe both sides of the plate glass clean before you leave it on the stage.

5.5 If there are something out of control about the microscope, don't turn it violently or take it apart. In order to avoid breaking it just tell the teacher.

5.6 Clean the microscope carefully after each experiment. You should have a high sense of responsibility.

5.7 Cover the microscope and take it back to the cupboard when it's not in use. Place a sack of silica drier in the cupboard to prevent the microscope from moisture.

第三节 基本实验技术

药用植物学实验中需要掌握的基本实验技术主要有：显微镜操作使用技术、生物绘图技术、标本制作技术、切片技术、显微摄影技术、组织培养技术等。实验室操作时，要掌握徒手切片、临时装片、压片、涂片、永久玻片制作，简单的显微化学测定等实验技术。现在分别加以介绍，有些实验技术在具体的实验中介绍。

一、徒手切片法

徒手切片法是用刀片把新鲜的或预先固定好的或软化的材料切成薄片，不染色或简单染色，用水封片后作临时观察，必要时也可制成永久性玻片标本。本法简便、迅速，能够观察到植物组织的生活状态，非常实用，但不易切薄切全，切片厚薄不均，不能做成连续切片。徒手切片的方法如下。

（1）材料先切成适当的段块，切片断面以不超过3～5mm^2，材料长度约2～3cm为宜，便于手持进行切片。软而薄的材料，如叶片可用胡萝卜根或马铃薯块茎或硬泡沫塑料等，夹住材料再一起切片。有的叶片可卷成筒状再切。坚硬材料可用水煮，软化后再切片。

（2）切片时，用左手拇指和食指夹住材料，并用中指拖住，并使其稍突出于手指之上，以右手执刀片，平放在左手的食指上，刀口向内，且与材料断面平行，移动右臂使刀口自左前方向右后方滑行切片，注意切勿来回拉锯，并在切片过程中应用水润湿材料的切面和刀面，以免切面破损。如此连续操作，将切下的薄片用湿毛笔轻移于盛水的培养皿中。再用毛笔挑选最薄的切片，取出放在玻片上，制成临时装片观察。也可用0.1%番红水溶液染细胞核和木质化、栓质化的细胞壁，以区分细胞核与细胞质，木质部（主要是导管）和韧皮部。

二、临时装片法

临时装片法是用少量材料，如薄的表皮、切成的薄片或粉末等，置在玻片上的水滴中，加盖盖玻片制成玻片标本，或选用甘油醋酸试液，水合氯醛试液等处理后观察。

加盖玻片时应注意先用镊子轻轻夹住盖玻片，使其边缘与材料边缘水滴的边缘接触，然后慢慢放下，放平盖玻片，使盖玻片下的空气逐渐被水挤出而不产生气泡，以免影响观察。

三、滑走切片法

滑走切片法是用滑走切片机进行切片，此法适合切制木材、木质的根茎等坚硬材料。滑走切片机能按需要调节切片的厚度，所切切片较完整，但不能做连续的切片。

四、解离组织法

解离组织法是用化学试剂，使细胞的胞间层溶解，使细胞彼此分离，以便观察不同组织的细胞形态特征。解离组织法常用于木质化程度较高的组织，一般来说，幼嫩组织不宜进行离析。常用的解离试剂有硝铬酸、氯酸钾和硝酸、氢氧化钾等。在解离前，应先将样品切成宽或厚约2mm的小条或片。

1. 硝铬酸解离 置材料于试管中，加10%硝酸与10%铬酸的等量混合液适量（约5ml），放置1天左右，或加温（30～45℃），以缩短浸渍时间，至用玻棒挤压能离散为止，倾去酸液，加水洗涤后，取出少量置载玻片上，用解剖针撕开，以稀甘油装片观察。

2. 氯酸钾和硝酸解离 置材料于试管中，加50%硝酸（约5ml）及氯酸钾少量，缓缓加热，待产生的气泡渐少时，再及时加入氯酸钾少量，以维持气泡稳定的发生，至用玻棒挤压能离散为止，倾去酸液，加水洗涤后，照硝铬酸解离法操作。此法和硝铬酸法适用于质地较坚硬，木化组织多的样品。

3. 氢氧化钾解离 置材料于试管中，加5%氢氧化钾溶液适量（约5ml），加热至用玻棒挤压能离散为止，倾去碱液，加水洗涤后，取出少量置载玻片上，用解剖针撕开，以稀甘油装置观察。此法适用于含薄壁组织多，木化组织少的样品。

五、石蜡切片法

石蜡切片法是制作永久性玻片标本常用的方法，其原理是对不易切的薄而均匀的样品，利用石蜡渗入植物组织中，使组织在石蜡的支持下，用旋转切片机进行切片，然后将切片中的石蜡除去，制作大量薄而均匀的连续切片标本，长期保存应用。石蜡切片制作步骤复杂，操作精细，时间长，不可能在一般实验中学习和掌握。下面简要介绍其制作步骤。

1. 取材和固定 将新鲜材料切成约5mm长和宽的小块，放入盛有F. A. A固定液的带橡皮塞的平底试管中，浸泡12～24小时，以保持细胞组织的细微构造。干燥材料不需固定，可用水或甘油浸软即行脱水。

2. 冲洗和脱水 用50%或70%乙醇浸泡及冲洗，其浓度应与固定液相近，然后逐渐提高乙醇的浓度，一般为70%、85%、95%和纯乙醇各等级。把材料中的水分脱除干净，每级处理时间1～3小时。

3. 透明 用50%的二甲苯纯乙醇溶液浸泡2～3小时后，换纯二甲苯处理，每次1～2小时，共处理两次，以除净材料中的乙醇，并使材料组织透明为止。

4. 浸蜡 取熔点52～54℃石蜡片少许，轻轻放入上述已透明材料的二甲苯溶液中，使石蜡逐渐溶解渗入材料中，并不断加入蜡片，直至石蜡饱和为止。

5. 烘蜡　将上述浸蜡至饱和的玻管放入恒温箱内，将温度逐渐升高，直至高出原石蜡熔点1～2℃，使二甲苯慢慢挥发，约2小时后，将材料移入盛有已融化的小烧杯中，每隔2～4小时换一次蜡，约2～3次后即可除尽二甲苯，使石蜡完全渗入材料中。

6. 包埋　将材料和石蜡液一起倒入小纸盒或包埋器中，并根据需要把材料按一定的间隔调整好位置，使之凝固（可放入凉水盆中），这样包埋有材料的蜡块，可长期保存供切片用。

7. 切片　将包埋好材料的蜡块切成适当大小，用烧热的蜡铲把蜡块粘固在小木块上，修正蜡块后，固定在旋转切片机上，切成连续的蜡带，其厚度通常为8～15μm。

8. 粘片　取洁净的载玻片，涂少许明胶粘贴剂，然后在其上加数滴蒸馏水，再将镜检合格的蜡片，以光面向下水平放水面，在35～45℃的烫板上，使蜡片展平，在常温或烘箱中干燥。

9. 脱蜡　将粘有蜡片的载玻片浸于纯二甲苯中，约10min，使石蜡完全溶去，以便染色。

10. 染色制片　常用番红和固绿二重染色。将溶去石蜡的切片，移入1/2二甲苯和1/2纯乙醇混合液中，再分别经95%、85%、70%乙醇处理，每级约2min，即可移入番红乙醇溶液，如用番红水溶液，则需经50%、30%、10%乙醇再行染色，染色时间2～24h，番红染色后，经过水洗，分级乙醇脱水，再行固绿复染，然后再经脱水、透明，即可用加拿大树胶或中性光学树脂加盖玻片封藏，贴标签即成永久玻片标本。

六、绘图技术

绘图是药用植物学实验报告的重要内容之一，绘图比文字记录更生动具体，能加深对植物形态和结构特征的理解，是学习植物形态和解剖时必须掌握的基本技能。植物绘图和美术绘图不同，其具体要求如下。

（1）首先要注意科学性和准确性，生物绘图不能夸张，应严格按照标本或显微镜下的影像写生。要认真观察要画的标本或切片等，选择正常的典型材料或区域影像，正确理解各部分特征，才能保证所绘图的科学性和准确性。

（2）布局要合理，画图前，应按实验要求的绘图数量和内容，在图纸上首先安排好构图的位置和相关部分的比例，并留出书写图题和注字的地方。在宏观上把握整张纸的内容安排，力求平衡，稳定和美观。

（3）先绘好草图，即用HB铅笔轻轻在图纸上勾画出图形的轮廓，然后用2H铅笔描出与观察对象相吻合的线条。线条要粗细均匀，光滑清晰，接头处无分叉，切忌重复描绘。

（4）植物图常用圆点的疏密表示明暗和颜色的深浅，点应圆而整齐，大小均匀，切忌用涂抹阴影的方法代替圆点。对绘制植物器官的显微结构简图，常用一些不同的划线方式和符号来表示各种组织的排列和分布（图1－6）。

（5）绘图纸要保持清洁，图注一律用铅笔正楷书写，并用平行线引出在图的右侧。

（6）实验题目写在图纸的上方，图题和所用材料的名称和部位写在图的下方。并注明放大倍数。

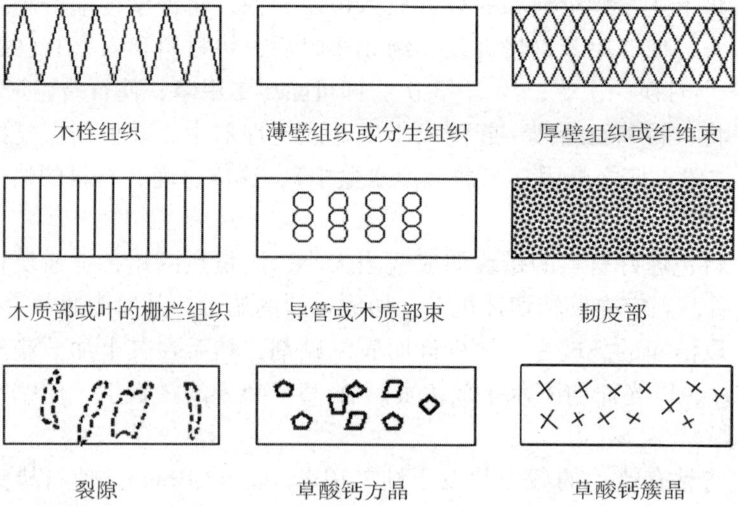

图 1－6 植物组织简图常用表示方法

七、简单的显微化学测定

显微化学方法是应用化学药剂处理植物的组织细胞，使其中某些微量的物质发生化学变化。从而产生特殊的染色反应，并通过显微镜来鉴定这些物质的性质及其分布状态的方法。其种类很多，下面仅介绍细胞壁的木质素成分和细胞中主要的三种贮藏物质的显微化学方法。

1. 淀粉的鉴定 淀粉是植物体中主要的贮藏物质，它们在不同植物细胞中形成各种不同形状的颗粒，当稀释的碘－碘化钾溶液与淀粉作用时，形成碘化淀粉，呈蓝色的特殊反应。所以用碘液测试淀粉已成为最常用的方法（配方见附录一）。但需注意如果碘液过浓，会使碘化淀粉变黑，反而不利于淀粉粒层纹及脐点的观察。

2. 蛋白质（糊粉粒）的鉴定 蛋白质是复杂的胶体，细胞内贮藏的蛋白质是没有生命的，呈比较稳定的状态，有无定形的、结晶状的或成为固定形态的糊粉粒。糊粉粒是植物细胞中贮藏蛋白质的主要形式。测试蛋白质常用的方法也是用碘－碘化钾溶液，但浓度较大效果才好（配方见附录一）。当碘液与细胞中的蛋白质作用时，呈黄色反应。在显微镜下观察，可见黄色的颗粒状的糊粉粒。注意在进行这种蛋白质鉴定工作之前，须用乙醇将材料进行处理，即在植物切片材料上滴加 95% 的乙醇，首先把材料中的脂肪溶解掉，以保证蛋白质颜色反应的正确性，并能看清糊粉粒的结构。

3. 脂肪和油滴的鉴定 脂肪和油滴也是植物细胞贮藏的主要营养物质之一。脂肪在常温下为固体形态，油滴则成液体状态，均不溶于水。

常用的显示脂肪的显微化学方法是苏丹Ⅲ的乙醇溶液染色（配方见附录一），呈橘红色，但近年已多用苏丹Ⅳ的丙酮溶液代替，其染色效果比前者稍红和明显。但这不是专一的组化反应，苏丹Ⅲ、Ⅳ均能使树脂、挥发油、角质和栓质染色。在鉴定过程中，为了效果明显，可稍加热以促进其反应。

4. 木质素的测定 木质素是芳香族的化合物，在细胞壁中一般呈复合状态。用盐酸和间苯三酚先后处理植物材料，是细胞壁木质素成分的鉴别方法。根据颜色反应的深浅能显示细胞壁中木质化的程度。

鉴别时，一般要取新鲜植物材料的切片，先加40%盐酸1~2滴，约3~5min后，待材料被盐酸浸透，再加5%间苯三酚的乙醇溶液（配方见附录一），当间苯三酚与细胞壁中的木质素相遇时，即发生樱红色或紫红色的反应。导管、管胞、纤维和石细胞等细胞壁中木质素成分丰富，因此它们的颜色反应十分典型。加盐酸的作用是由于间苯三酚需在酸性环境中才能发生上述反应。间苯三酚为白色粉末，易氧化变性，若已呈灰褐色或溶液已发黄，往往无效。

Section 3 Elementary Experiment Technology

There are some elementary experimental technologies you should grasp in pharmaceutical botany, they include: operation of microscope, biological drawing, making medicinal plant specimen, slice cutting technology, micro - photograph, tissue culture technology and so on. In the laboratory, free - hand sectioning, making temporary section, press slice, smear slice, permanent slice and simple measurement through microscope and chemical agent also needed to be grasped. Now we introduce some technologies here, other technologies will be studied in the specific experiments.

1. Method of free - hand sectioning

Method of free - hand sectioning is that of cutting fresh materials or fixed materials before or softened materials into slice with razor blade, without dyeing, or simple dyeing, and observes temporarily after slices are covered with water. It can also be made into the permanent specimen when it is necessary. This method is convenient, swift, practical, which can observe the live situation of plant tissues. However it is not easy to cut the slices fully and thinly, sometimes the thickness of slice is not even, sometimes it can not be made into continuous slices. Procedure for free - hand sectioning is as follows.

1.1 At first, cut the materials into the pieces or segments, which the area is less than 3 - 5mm^2 and the length of it is less than 2 - 3cm, that is fit to hold it and cut the slice. To soft and thin materials, leaves for example, you can clasp it with root of carrot, tuber of potato or hard foam and so on. Some kind of leaf may be curved into cylinder then to cut the slice, to sturdy material, cut the slice after it is softened by boiling with water.

1.2 When cutting slice, clasp the material with thumb and forefinger of the left hand, and hold it with the middle finger to make sure that the material is higher than fingers. Hold the razor blade with the right hand, after that put the razor blade on the forefinger of left hand with the edge of razor towards operator and at the same time make sure that the edge is parallel with the section of material. Then move the right arm to move the razor blade slide from the left to right, and not be locked in a seesaw. During sliding, you should keep the section of material and the edge of razor blade wet or it will damage the slice. Do this continuously, and put the slice just made into a basin full of water with a wet writing brush. Choose the thinnest slice and place it on the slide to make a temporary specimen, or else, dye nucleus and suberized, lignified cell wall with safranine to differentiate nucleus from cytoplasm, xylem (main is vessel) from phloem.

2. Method of making temporary section

The method of making temporary section is to place a little materials, such as a thin epidermis, thin slice or powder of samples into the water drop on the slide, then put the cover glass on the slide, and that is the specimen. You can also choose glycerine acetic acid solvent, chloral hydrate and other reagent to make temporary section.

When placing a cover glass, please pay attention clamping the cover glass slightly with forceps to make sure that its margin connects with the edge of water drop which is on the edge of material, then put the cover glass down slowly to assure the air behind the cover glass can be eliminated without air bubble, so as to avoid obstructing observation.

3. Method of sliding slice

Method of sliding slice is consist of cutting with a sliding microtome. This method is suitable to cut sturdy material, such as wood, lignified root or lignified stem and so on. Sliding microtome can adjust the thickness of slice following operator's request, and the slice is more integrated than other methods. However the machine can not do it continuously.

4. Method of dissociation tissue

The method of dissociation tissue is a method using chemical agent to dissolve intercellular layer to make cellsseparate, so as to observe features of cell shape in different tissues. This method often be used for the tissue that is lignified strongly, and is not suitable for the dissociation of very young tissue. Commonly used dissolving agent includes potassium hydrate, nitric acid and chromic acid, potassium chlorate and nitric acid, ect. Cut the samples into 2mm slice before dissociation.

4.1 Nitric acid and chromic acid dissociation Put the materials into tubes, add a mixed solution with the same volume of 10% nitric acid and 10% chromic acid, display about 24 hours, or heat up (30–45℃) to shorten infiltration time, till its dissociation by compression with glass rod, drop out the acid solvent, toilet with water, take some out onto slide, tear it dissecting needle, treat it with dilute glycerin, cover with a cover glass then to observe.

4.2 Potassium chlorate and nitric acid dissociation Put the materials into tubes, add some 50% nitric acid (about 50ml) and some potassium chlorate and heat gradually. In the meantime add some potassium chlorate when bubbles leave off to maintain bubbles' numbers, till its dissociation by compression with glass rod, drop out the acid solvent, toilet with water, then treat it as 4.1

This method is fit for the hard samples containing more xylem tissues.

4.3 Potassium hydrate dissociation Put the materials into tubes, add some 5% potassium hydrate (about 5ml) into it, heating till its dissociation by compression with glass rod, drop out the basic solvent, wash with water, take some out onto the slide, tear with dissecting needle, treat it with dilute glycerin then to observe. This method is fit for the samples containing many parenchyma and little xylem tissues.

5. Method of paraffin wax slice

Paraffin wax slice is a common method to make permanent slides. It is concerned with samples, with which is difficult to cut the slice thin and equitable. In this situation, we use (waxes) that can permeate into plant issues. The issues being hold of wax is cut into slice by rotary microtome. Then eliminate the wax in the slice and make a great many thin and equitable consistant slices to preserve to use. Paraffin wax slice is difficult to make with complex processes, concise manipulation and long period, so it can hardly be learned and mastered in the common short time experiment. We only introduce it briefly as follows.

5.1 Fixation Cut the fresh material into blocks, which the length and width are about 5mm, then soak the blocks in F. A. A. (Formalin Aceto - Alcohol Solution) fixatives that is in a flat - bottom test - tube with rubber stopper for 12 - 24 hours to protect the structures of cell tissues. For dry materials, it need not to be fixed with F. A. A., just soften it with water or glycerine, then to dehydrate.

5.2 Rinsing and dehydration Rinsing is to soak the tissues with 50% or 70% alcohol. At first the concentration of the alcohol should be similar to that of the fixative, then increase the concentration of the alcohol by turns 70%, 85%, 95% and anhydrous alcohol. Each concentration should soak for 1 - 3 hours so as to sweep out the water of the material.

5.3 Transparentizing Leave such material in equal parts xylene and absolute alcohol at intervals of 2 - 3 hours, then rinse 2 - 4 hours in two change of absolute xylene to clear ethyl alcohol in the material and stop after the tissues has become transparent.

5.4 Infiltration with paraffin Cast a little paraffin melted point ranging from 52℃ to 54℃ into xylene solvent, paraffin gradually dissolves and penetrate into the tissues. Continue the addition of melted wax until the solvent is saturated with paraffin at this temperature.

5.5 Bake Transfer the tube containing above saturated solvent into the oven, improve temperature gradually until the temperature is beyond the paraffin's melting point 1 - 2℃ to make xylene evaporate. After about 2 hours, transfer the material into a beaker containing melted paraffin, embed 2 - 4 hours in paraffin in two or three changes of paraffin, after that the xylene and the paraffin diffuses downward into the tissues can be swept out completely.

5.6 Embedding Pour the paraffin containing the tissues into the mold or paper box, arrange the pieces required at suitable intervals to harden the paraffin. Such paraffin block containing the material can be stored permanently and used for sectioning.

5.7 Sectioning Cut the paraffin blocks containing the embedded samples into a suitable length, then stick the paraffin block to a wood block with a hoted shovel, fix the paraffin block on the rotary sliding microtome, cut it into serial sections. Usually the thickness of the section is about 8 - 15μm.

5.8 Fasten Cast gelatin adhesive onto a clean slide and add several drops of water, place the qualified sections examined by the microscope on the water drops, be sure that the section's flat side up on. Then put the slide on a heated plate with the temperature 35 - 45℃ to make the paraffin flatten out, and dry at room temperature or in oven.

5.9 Dewax Soak such slide fasten with paraffin section into absolute xylene for about ten minutes to dissolve paraffin completely, so as to stain easily.

5.10 Stain Safranine and fast green are usually used in two - times staining. At first put the section replaced paraffin into the solution with equal volumes of xylene and absolute alcohol, then to make it through decreased grades of ethyl alcohol, such as 95%, 85%, 70%, use 2 minutes intervals decrease to next grade. If using aqueous safranin to stain, you should still soak the section into the additional solvents that is made through the decreased grades of ethyl alcohol, 50%, 30%, 10%. Staining in safranine should be sustained for about 2 - 24 hours. After staining in safranine, run down to water, dehydrate with grades of ethyl alcohol, restain with fast green, then to dehydrate again and transparentize, mount with Canadian gum or neutral optical resin and cover glass, stick the label to make a permanent slide.

6. Technology and methods in experimental drawing

Experimental drawing is an important part of an experimental report, it can deepen your understanding on a plant's morphology and structure more vividly and correctly than linguistic description. Also, it is an elementary and necessary skill that you must master in learning of dissection. Being different from artistic painting, experimental drawing here are some orders.

6.1 To make your drawing scientific and accurate. When drawing, you should strictly obey the sample and image under microscope, watch your specimens carefully, choose the most classic sample or image and grasp the characters of each section correctly.

6.2 Make a frame. Arrange all the plottings and each ratio according to the experiment direction, which will tell you how many pictures and what contents should be included in your paper. In addition, leave enough space for topic and annotation for every drawing.

6.3 Besides, you should take a little more trouble in sketching. Firstly, draw the outline of the figure with a HB pencil, and then depict it meticulously based on the appearances you have seen under your microscope with a 2H pencil. The lines should be even, smooth, clear, branching is not allowed, so please remember by heart: don't trace the line repeatedly.

6.4 Rather than shadow, round regular spots, which have different density, are often used to express the light intensity and colour difference in a plant's drawing. When making an abbreviated one, you need more standardized manners to show the organization of microstructure (Figure 1 - 6).

6.5 Keep your paper clean. In penciled regular script, the annotation should be placed on the right side of the drawing, being connected with different position inparallels.

6.6 Put the title on the head of the paper. Take notes on a specific topic, plant name, dissection parts as well as magnifications under each drawing.

7. Simple measurement through microscope and chemical agent

Microscopic and chemical measurement is to handle botanical histocyte with chemical agent, and make some traces of substance occur chemical change, thus to produce the special staining reaction, then to identify the property and distribution of these substance. There are many kinds of methods, here we only introduce microscopic and chemical methods of the lignin in the cell wall and the three main storage materials in the cell.

7.1 Identification of starch Starch is the main storage material in the vegetal body and the shape of the starch from different plant varies. Starch will be colored blue with the dilute solution of Iodine – Iodination kalium, forming iodination starch. So Iodine Water is commonly used in testing starch. (for recipes see appendix one). But you should pay attention to that Iodine Water is not too thick to gain the observation of layers of starch granule and hilum as a result of black iodination starch.

7.2 Identification of proteins (aleurone grain) Protein is a kind of complex colloid, and some stored in the cell is not vital but much stable. The form of the storage protein includes amorphous, crystalline or morphological aleurone grain. Aleurone grain is the main form of storage protein in the plant. Iodine – Iodination kalium is also commonly used to test protein, but it is better in case of higher concentration. Iodine Solution stains the protein in the cell yellow. Yellow and granulose aleurone grains can be observed through the microscope. Please pay attention that before identifying protein, you should drop 95% alcohol to the material of botanical slide to dissolve the fat of material, and so guarantee the accuracy of staining reaction of the protein and clearly observe the structure of aleurone grain.

7.3 Identification of fats and oils Fat and oils are alsothe main nutrient substances stored in the plant cell. Fats occur in solid masses at the usual temperature, but oils occur as droplets in the cell and both are insoluble in water.

The alcohol solution of Sudan III is often used to test fats (for recipes see Appendix one) . It is mainly responsible for orange – red coloration, but over the last years, Acetone Solution of Sudan IV is often used instead and stains fat a little redder and more obvious. But it is not a special test for fats, both Sudan III and Sudan IV can stain resin, volatile oil and cutin. In the course of the identification, in order to gain obvious effect, it is accessible to gain a little heat.

7.4 Identificationof lignin Lignin is an aromatic compound and creats compounds in the cell wall. The method of identification of lignin is to handle plant material with Hydrochloric Acid and Phloroglucinol in turn. The lignified level of cell wall can be tested through the degree of staining reaction.

When identifying, take fresh slices of plant material to glass slide, first drop 1 – 2 droplets of 40% Hydrochloric Acid, and then Alcoholic solution of 5% Phloroglucin (recipe sees Appendix one) and lignin in the cell wall is colored vinaceous with Phloroglucin. The lignin is abundant in the cell wall of vessel, tracheid, fiber and stone cell, so this staining reaction is quite typical in these structures. Dropping Hydrochloric Acid is because Phloroglucin can occur above - mentioned reaction only in acid environment.

Phloroglucin is white powder and readily oxidized, so if it has been dun or the solution has become xanthous, usually invalid.

第四节　药用植物标本的制作方法

一、蜡叶标本的制作

所谓蜡叶标本，就是把采集到的新鲜植物全株或一部分，经压制去掉水分后的干燥植株标本，也叫“压制标本”。此法对种子植物、蕨类及苔藓类植物均适用。其制作方法大致如下。

1. 压制前的准备工作　宜选用形态完好、大小适宜的鲜标本，将多余的枝叶进行修剪，去掉泥土等杂物。细小的种子及花粉宜单独装入小袋中，与该植物标本放在一起。粗大的地下部分宜切割一小部分，或与其地上部分的标本编成同一号码后另行干燥。

2. 做好野外记录　要按野外记录所列各项认真填写，特别是压制后容易改变的特征：如花、果实的颜色，叶片上的白粉，容易脱落的毛茸等要详细记录。

3. 挂好牌号　按牌号上所列各项填写后，拴在植物标本上。号码的号牌必须与野外记录号牌一致。

4. 压制蜡叶标本　压制蜡叶标本时，首先将标本架子或标本厚夹板的一片放平，然后在上面放几张吸水纸；吸水纸上面放一张夹标本的纸作为内页，将经过挑选并拴好号牌的新鲜植物与野外记录一起夹在里面，其上面再放几张吸水纸，如此反复操作。最后将另一标本夹（或夹板）盖上，用绳子捆紧，放在通风良好处干燥。干后需换一次吸水纸：即用干燥吸水纸把已潮湿的吸水纸一次替换下来。在换吸水纸时夹标本的纸不要更换，否则容易损坏标本。遇阴雨天，每天至少要换两次吸水纸。如上所述，每天换吸水纸，直到标本完全干燥为止。

5. 标本的消毒　野外直接采集的标本或外单位交换来的标本往往带有害虫卵或霉菌孢子，故在标本入柜之前必须进行消毒，消毒方法如下：配制氯化高汞（$HgCl_2$，升汞）的千分之二至千分之五的乙醇溶液适量放在搪瓷盘内，将标本浸透在升汞的乙醇溶液中静置5~10min，用竹夹夹取放在干的吸水纸中，乙醇挥发，而$HgCl_2$留在标本上，标本干后即可上台纸。$HgCl_2$有剧毒，操作时应注意房间通风，切忌用手直接操作，要带胶皮手套和口罩，操作后要洗手以免中毒。剩余消毒液要妥善保管。目前有用冷冻处理来进行标本的消毒。

6. 标本的装订　已经消毒的标本用毛笔将胶水（最好用植物胶）刷在标本背面，花的部分不必上胶，以便解剖观察花的各部形态，然后移贴台纸上，台纸的规格为39cm×27cm的白纸板。可用约40cm×30cm的厚卡片纸稍加压力，放置待干时，应注意在左上角和右下角分别留出贴野外记录签和定名签的位置，然后用纸条或线将叶片和植物粗壮部分穿钉固定在台纸上即成。

7. 标本的鉴定与存放 标本装订完毕后，接着要进行鉴定。首先鉴定到科，利用《中国高等植物分科检索表》进行检索分科，确定科别后，再利用《中国植物志》、《中国高等植物图鉴》、《东北植物检索表》等地方植物志进行分属和分种检索，最后确定种名。鉴定标本时，要利用放大镜和解剖镜对标本仔细观察，按照检索表的各条款项逐一核对，尤其注意花、果实、种子等繁殖器官的核对。

标本经过分科、分属和分种鉴定后，可将定名签贴在台纸右下角，野外记录签贴在左上角，最后可加贴一张薄而韧性强的封面衬纸，以免标本互相摩擦损坏，这样即成为完整的标本。随后将同种植物标本放在一起，用种夹夹起，可用牛皮纸按台纸大小制作并在种夹外注明该种植物学名，按科属顺序放入标本柜中密闭保存，柜中可放入一些樟脑球防虫，整个标本室可用硫酰氟或溴代甲烷熏蒸消毒，但消毒后要打开窗户通风数日方可进入。

二、浸制标本的制作

浸制标本是将新鲜植物浸泡在具有杀菌防腐作用的溶液里，借以较长时期保存标本的方法。制作时，将完整的植物或经过挑选和修剪的新鲜植物标本放进玻璃容器（通常用大型磨口玻璃标本瓶或磨口缸）中，然后加 10% 甲醛溶液或加 30% ~50% 的乙醇和少许甘油，使之浸没植物标本，最后加盖封闭保存。在玻璃容器外面贴上标签。浸制标本适用于新鲜花朵、各种地下茎、大型肉质果实及多汁的植物的长期保存。对于某些藻类、真菌等，应将每种植物分别浸制并装入不同的容器内，以免孢子相互混淆。浸制标本使用的固定液还有 F. A. A. 固定液、5% ~6% 的甲醛水溶液、60% ~70% 乙醇水溶液、3% ~5% 冰醋酸水溶液等。

Section 4 Method in Making Medicinal Plant Specimen

1. Make Herbarium sheet

A herbarium sheet is a kind of dry plant sample. People collect fresh plants from outdoors, take a total plant or a part and get rid of the water by pressing. This is a simple case, and such a sheet is also called "a pressing sample". The method is widely used in the conversation of spermatophyte, fern, and lichen. Specific methods as follows.

1.1 Choose integral and proper plants from your collections remove surplus leaves and branches and get rid of the mud. Tiny seeds and pollen should be saved into small packages and put together with the plant. If the hypogeal part is too large, you had better remove a small part of it or sign the same number with its overground part and dry it additionally.

1.2 Take a careful record according to the items given by the outdoor directions, especially about the changeful characters (eg. The color of flowers and fruits, white power of leaves, easy – shedded fuzz and so on).

1.3 Hang the sign cards on different plants after filling in all items. The sign number should be kept in accordance with the outdoor record numbers.

1.4 Press the herbarium sheet Firstly lay down the sample shelf or plywood and put several pieces of water – absorbing paper on it; secondly put another piece of paper on the water – absorbing paper to sandwich your sample which has already been chosen and numbered together with the outdoor records; then, you need to put some water – absorbing paper on it again. Repeat this process again and again. At last, cover the whole accumulation with another shelf or plywood and bundle them tightly. Put it into the airy condition for drying. When it seems dry, you should substitute the old wet paper with the new dry one. But do not change the sample – sandwiching paper. Otherwise the samples may be spoiled. If it is gloomy, you should change the paper twice a day at least. As mentioned above, change the water – absorbing paper everyday until the samples are absolutely dry.

1.5 Sample disinfection. There are always harmful worm eggs and mildew in outdoor plants or quid pro quo with other units. So before putting them into the sample cupboard, the samples must bedisinfected. The method is: pour alcohol solution and 2‰ – 5‰ mercuric chloride onto a porcelain plate, let the sample macerate in it for 5 ~ 10 minutes, then clip it out and put it down to the dry waterabsorbing paper, the alcohol will be volatilized, leaving only mercuric chloride. Press it until it is dry, and then move it to the backer, which is a piece of very hard paper as a holder. Mercuric chloride is virulent, so during disinfection, you should keep good ventilation, do not handle it directly with your hands. Instead, wear plastic gloves and gauze mask, wash your hand in time after accomplishing it. Besides, the remaining disinfectant should be reserved carefully. Recently, the cold methods to disinfection is also be used.

1.6 Sample biding. You can add glue by using a writing brush (mucilage is better) and swearing it on the back of the sample, be careful not on the flowers to make sure your dissection and observation of flowers more easier. Then move it to the backer which is usually a white cardboardpaper by 39cm × 27cm and start to press it, at this point, a thick card by 40cm × 30cm or so may be useful. Leave enough space to affix your outdoor records and name lots. Finally, fix the leaves and strong parts with scrip or thread.

1.7 Sample identification and conservation. When to identify the sample, here are some useful references books for you: *Higher plant family index of China* for the family identification, and *Flora of China*, *Higher plant picture of China*, *Plant index of northeast of China* and other regional floras for genus, and confirm its name. In this process, you need a careful observation with a magnification and anatomic lens, paying more attention to flowers, fruits and seeds according to the references above.

After identifying a sample's family, category, species, you can adhere the name lots to the baker under right corner, and outdoor records up left corner. Then cover the sample with a thin and flexible paper to protect the sample against friction. Thus a single herbarium sheet is finished. Next, put the same species together in a nip, you can use kraft, which have the same size with backer to record the scientific name of each species, and then put these nips into cupboards according to their family, and genus orders for sealed conservation. Besides, you can use some camphor to keep insects away. The whole sample - room can be disinfected by sulfuryl fluoride or bromethane steam. After disinfecting the sample - room, it should be kept airy for several days before entering.

2. Make soaking sample

In order to conserve the samples for a longer time, you need to keep the fresh plants into the antiseptic solution, this is called "soaking sample" . At first, put the total plant or only a part selected and pruned into a glass container (usually using ground glass bottles or jars) . Then add 10% formalin or 30% to 50% alcohol and a little glycerine to make sure the sample is steeped totally. Finally, lid the container for sealing and label it outside. This method is often used in the long conservation of fresh flowers, rhizomes, big and fleshy fruits as well as juicy plants. Besides, sometimes same kind of plants (eg. alga, fungi…) should be saved in different containers to avoid the spore mix with each other. Also the F. A. A solution, 5% - 6% formaldehyde, 60% - 70% alcohol, and 3% - 5% acetic acid can be used in soaking samples.

第二章 实验内容

Chapter 2 Experiment Content

实验一 植物细胞的构造及后含物

【目的要求】

1. 掌握植物细胞的基本构造。
2. 了解分布于植物细胞中的主要后含物的种类及其形态特征。
3. 学习临时装片法及绘制植物细胞图的基本技术。

【实验材料】

洋葱鳞茎、马铃薯块茎、蓖麻种子、桔梗根、大黄粉末、半夏粉末、甘草或黄柏粉末、印度橡皮树叶或无花果叶的横切片、苍术根的横切片。

【仪器用品】

显微镜、载玻片、盖玻片、解剖用具、纱布、酒精灯、培养皿、擦镜纸、吸水纸、蒸馏水、稀碘液、水合氯醛试液、稀甘油、95%乙醇和75%乙醇。

【内容与方法】

一、植物细胞的基本构造

1. 制作洋葱表皮的临时装片 从洋葱鳞茎上剥下一片肉质鳞叶，用刀片在其内表皮上划一个小井字，使井字中央切割的小块长宽成3~5mm。用镊子撕下井字中央这小块透明的内表皮，按临时装片法（参见第一章），置于载玻片上预先加好的水滴中，用镊子将其展平，然后将盖玻片沿水滴一侧慢慢盖下，防止产生气泡，用吸水纸沿盖玻片一侧吸掉多余的水。

2. 观察植物细胞的构造 在阅读第一章有关显微镜的构造和使用方法后，将临时装片放在低倍镜下观察，注意洋葱鳞叶内表皮为一层细胞，细胞为长方形或扁砖状，排列紧密，没有细胞间隙。移动装片，选择几个较清楚的细胞置于视野中央，换用高倍镜再仔细观察，注意识别下列各部分结构（附图1-1）。

（1）细胞壁 为植物细胞所特有。位于细胞的最外层，比较透明。调解细调焦螺旋和虹彩光圈时，可见这层细胞壁实际为三层，其两侧为相邻两个细胞的细胞壁，中间是两个细胞所共有的中胶层（胞间层）。

（2）细胞质 为无色透明的胶状物，紧贴在细胞壁以内，被中央大液泡挤成一薄层，仅细胞两端较明显。

（3）细胞核 为扁圆球形的小球体，浸埋在细胞质中，贴近细胞侧壁的只能见其窄面，贴在细胞上下壁的可见其宽面。仔细观察，可见包在细胞核外的核膜，细胞核内有1～3个发亮的小颗粒，即核仁。

（4）液泡 位于细胞中央，占细胞体积的大部分。由于液泡内充满细胞液，所以比细胞质更透明。为了观察清楚，可取下制片，小心由盖玻片一侧滴加一滴稀碘液，由盖玻片另一侧用吸水纸吸，使稀碘液渗入盖玻片下，几分钟后观察，可见细胞质被染成浅黄色，细胞核被染成较深的黄色。

二、植物细胞的后含物

1. 淀粉粒 割取马铃薯块茎一小块，用刀片在断面刮取少许“白浆”，置载玻片上，加一滴蒸馏水制成临时装片，先在低倍镜下观察淀粉粒，注意其形状。再换用高倍镜观察，注意其脐点和层纹，分辨出单粒、复粒和半复粒（附图1－2）。然后再观察半夏粉末，注意其和马铃薯淀粉粒有何不同。在盖玻片一侧加一滴稀碘液染色，观察有何颜色变化。

2. 结晶体

（1）草酸钙结晶

①簇晶 取大黄粉末少许，置载玻片上，滴加水合氯醛试液1～2滴。在酒精灯上慢慢加热进行透化，注意不要蒸干，可加添新的试剂，并用滤纸吸去已带色的多余试剂，至材料颜色变浅而透明时，停止处理，加稀甘油一滴，盖上盖玻片，拭净其周围的试剂，镜检，可见多数大型、星状的草酸钙簇晶（附图1－3）。

②针晶 取半夏粉末少许，置载玻片上，如上法透化后，镜检，可见散在或成束的针状草酸钙晶体（附图1－4）。

③方晶 取黄柏或甘草粉末少许，如上法透化后，镜检，可见在细长的成束的纤维周围的薄壁细胞内，含有方形或长方形的草酸钙晶体（附图1－5），这种结构称为晶鞘纤维或晶纤维。

（2）碳酸钙结晶 观察印度橡皮树叶或无花果叶的横切片，注意在近叶面的表皮细胞的大型细胞中，有葡萄状的结晶体附着在细胞壁增生的棒状物上，悬挂在细胞腔中，形似钟乳石，即碳酸钙结晶，又称钟乳体（附图1－6）。

3. 菊糖 取浸于95%乙醇中4～5天的桔梗根一段，按徒手切片法（参考第一章），直接用75%乙醇溶液制成临时装片，置显微镜下观察，可见在薄壁细胞中靠近细胞壁有呈球形或扇形，并有放射状纹理的菊糖结晶。

另取苍术根的横切片置显微镜下，区别二者的异同点。

4. 糊粉粒 取蓖麻种子的胚乳做徒手切片，放载玻片上。先滴几滴95%乙醇，以便将材料中的脂肪溶解掉，再加一滴稀碘液，封片。在高倍镜下观察，可见到被染成黄色的多边形蛋白质晶体，不被染色的含磷酸盐的球晶体及无定形的基质。

【实验报告】

1. 绘洋葱鳞叶的内表皮细胞图，注明细胞各部分名称。

2. 绘马铃薯、半夏块茎中淀粉粒的构造和类型图，注明各部位名称。

3. 绘各种草酸钙结晶的形态图。

【思考题】

1. 制作显微标本装片时，怎样才能防止气泡发生？

2. 为什么在显微镜下观察草酸钙结晶时需要透化？

3. 如何区分草酸钙结晶和碳酸钙结晶？

Experiment 1 Structure and Main Ergastic Substance of Plant Cells

Aim and demand

1. Grasp the basic structure of plant cells.

2. Understand the types and shapes of the main ergastic substances of the cell.

3. Study temporary slicing technique and the drawing skill of plant cells.

Experimental materials

Bulb of Yangcong (*Allium cepa* L.), tuber of Malingshu (*Solanum tuberosum* L.), castor bean of Bima (*Ricinus communis* L.), root of Jiegeng (*Platycodon grandiflorum* (Jacq.) A. DC.), powder of rhizome of Dahuang (*Rheum officinalis* Bill.), powder of tuber of Banxia (*Pinellia ternata* (Thunb.) Breit), powder of root of Gancao (*Glycyrrhiza uralensis* Fisch.) or stem peel of Huangbai (*Phellodendron amurense* Rupr.), slide of cross section of leaf of Induxiangpishu (*Ficus elastica* Roxb.) or Wuhuaguo (*Ficus carica* L.), cross section of root of Changzhu (*Atractylodes lancea* (Thunb.) DC.).

Instruments and appliances

Microscope, slide, cover glass, dissector, spirit lamp, culture dish, lens paper, filter paper, distilled water, thin iodine solution, chloral hydrate reagent, thin glycerine, 95% alcohol, 75% alcohol.

Content and procedures

1. Basic structure of plant cells

1.1 Make temporary slides for onion scale leaf Cut a slice of fleshy scale leaf from *Allium cepa* bulb, using forceps remove a transparent layer of the inner skin, cut into a piece 3～5mm long, following instructions for temporary slide making (refer to chapter 1), place sample onto an already prepared slide containing one drop of water. Smooth out the thin layer with forceps, then place another cover glass onto and slowly put down the cover glass to avoid bubbles. Lastly remove any excess water using absorbent paper.

1.2 Observe the structure of plant cells After reading the construction and usage of the microscope, place the prepared slide under the scope. Notice that the inner layer of the onion skin is comprised of a single layer of cells. The cells are rectangular in shape, closely packed, without much intercellular space. Shift the slide and focus on a few well looked cells, switch to a higher magnification, to notice the following (Figure 1 – 1).

1.2.1 The cell wall Cell walls are unique to plant cells. It surrounds the cell and appears clear. When adjust fine adjustment and iris diaphragm, three layers of the cell wall can be seen, two layers of cell walls with a layer of middle lamella.

1.2.2 The cytoplasm The cell is composed of a clear gelatinous substance, stuck closely to the inner cell wall, it is extruded into a thin layer by the central vacuoles, only the two end of the cell is clear.

1.2.3 Nucleus The nucleus is a oblate spheroidal small spheroid, buried inside the cell body. Only the narrow side can be seen from the sides of the cell, the wide sides can be seen from the top of bottom of the cell. Also, 1 ~ 3 shiny dots can be seen inside the nuclear membrane, that is nucleolus.

1.2.4 Vacuole Situated at the center of the cell, comprises most of the volume of the cell. Because the cell is filled with cell sap, the vacuole appears clear. In order to see the vacuole more clearly, carefully add a drop of thin iodine solution. Observe after a few minutes, pay attention that the cell now has a yellow tint and the nucleus is dark yellow.

2. Ergastic substances of plant cell

2.1 Starch Take a small piece oftuber of *Solanum tuberosum*, use a razor to scrape away part of the "white liquid", place on a slide and add a drop of distilled water to make a temporary slide. Firstly to observe under low powered microscope, note its shape, then switch to a higher powered microscope and observe, notice hilum and lamellae, distinguish simple starch grain, compound grain and half compound grain. Then to observe powder of tuber of *Pinellia ternata*, pay attention to the difference with potato starch grain (Figure 1 – 2). At last add a drop of iodine to the slide, to observe the change in color.

2.2 Crystals

2.2.1 Calcium oxalate crystals

2.2.1.1 Cluster crystal Take a little *Rheum officinalis* powder, put on the slide, add chloral hydrate reagent 1 ~ 2 drops. Heated on an alcohol lamp carefully not to let it dry up completely, new solution can be added if necessary, excess solution can be removed with absorbent paper. When sample has become clear, stop the treatment. Add a drop of thin glycerine, cover with a cover glass, wipe off the overflowed glycerine solution around the cover glass, then to observe under a microscope. You can find many large radial calcium oxalate cluster crystals (Figure 1 – 3).

2.2.1.2 Acicular crystal Take small amount of powder of tuber of *Pinellia ternata*, follow the process above to make it clear, then to observe. You can see the acicular crystal, which is scattered or aggregated into bundles (Figure 1 – 4).

2.2.1.3 Solitary Take little amount of powder of *Glycyrrhiza uralensis* root or *Phellodendron amurense* stem peel after transparentizing with chloral hydrate, to observe under a microscope. You can see that there are many square or rectangular calcium oxalate crystals in the parenchyma cell around the long fiber bundles. This structure is named crystal fiber. (Figure 1 – 5).

2.2.2 Calcium carbonate Observe cross section of leaf of *Ficus elastica* or *Ficus carica*, and notice that inside the large cells on the leaf surface there are grape shaped crystalline bodies adhered to a stick shape substance that comes from hyperplastic cell wall. The crystalline body hangs in the cell like stalactite, this is the calcium carbonate crystal, another name is cystolith. (Figure 1 – 6).

2.3 Inulins Take a little section of root of *Platycodon grandiflorum* that immersed in 95% alcohol for 4 – 5 days, follow the method of free – hand Sectioning (refer to Chapter 1) to make the temporary 75% alcohol slide, then to observe under a microscope. You can see that there are spheroidal or fan shaped inulins crystal near the cell wall of parenchyma cell, it has radial veins. Then to observe the cross section of root of Chinese atractylodes under a microscope to distinct it with the balloonflower.

2.4 Aleurone grain Take a castor bean of *Ricinus communis*, pick out its endosperm, then to use it make a free – hand section. You can place a few drops of 95% alcohol solution onto the slide to dissolve any lipids in the section sample. Then place a drop of iodine onto the slide, and cover a cover glass to observe under high powered microscope, yellow stained polygonal protein crystals should be visible. Those that are not stained are amorphous stroma and spherical crystals that contains phosphate.

Experimentreport

1. Draw a inner epidermis cell of *Allium cepa* scale leaf, note the name of each part of the cell.

2. Draw the structure and types of starches of tuber of *Solanum tuberosum* and tuber of *Pinellia ternata*, note the name of each part of the starch.

3. Draw the morphological figure of all types calcium oxalate crystals you have observed.

Questions

1. While making slides, how can you avoid the formation of bubbles?
2. Why must we make the solution clear when observing calcium oxalate crystal?
3. How do you distinguish calcium oxalate crystals and calcium carbonate crystals?

实验二 分生组织、保护组织与机械组织

【目的和要求】

1. 了解分生组织的形态特征。
2. 掌握保护组织的形态特征及其在植物体的分布部位。
3. 掌握机械组织的细胞形态和结构特征。

【实验材料】

常山叶、薄荷叶或西洋红（一串红）叶、曼陀罗叶、天竺葵叶、菘蓝叶或独行菜叶，白藓皮粉末、石榴皮粉末。

洋葱根尖纵切片、薄荷茎或南瓜茎横切片、甘草根横切片、接骨木茎横切片。

【仪器用品】

显微镜、解剖用具、酒精灯、水合氯醛试液、稀甘油、间苯三酚试液、盐酸。

【内容与方法】

一、分生组织

顶端分生组织：观察洋葱根尖纵切片，先从低倍镜下找到细胞最小、染色最深的圆锥形的根端生长锥，即根尖分生组织所在部位。可大概分成两部分：前者（最顶端）为一群最小最幼嫩的细胞，具有强烈持久的分裂能力，称为原分生组织，其细胞壁薄，细胞质浓，细胞核大，液泡很小而多，细胞排列整齐，没有细胞间隙。后一部分的细胞已有初步分化，其最外一层为原表皮，中央柱状部分为原形成层，在其与表皮之间的区域为基本分生组织。在生长锥的前方为帽状的根冠。

二、保护组织

1. 表皮 撕取薄荷叶的下表皮一小块，制成临时水装片，镜检，可见表皮由一层扁平薄壁细胞组成，细胞排列紧密，不含叶绿体，表面观其侧壁多呈波状，细胞相互嵌合，衔接坚牢。注意在表皮能否看到角质层、毛茸和气孔。

2. 毛茸

（1）非腺毛 分别撕取菘蓝叶和天竺葵叶的下表皮一小块，制成临时水装片，镜检，可见顶端尖狭的非腺毛（附图 2-1），注意其由几个细胞组成。

（2）腺毛

①观察薄荷叶或西洋红叶下表皮水装片，注意其腺毛有两种类型，一种具 1~2 个细胞的腺头和 1~2 个细胞的腺柄（附图 2-1），另一种为腺鳞，腺头由 6~8 个细胞组成，略成扁球形，排列在同一平面上，周围有角质层，其与腺头细胞之间贮有挥发油，腺柄极短（附图 2-2）。

②观察曼陀罗叶下表皮水装片，注意其腺毛的形态和结构。

3. 气孔

（1）平轴式气孔 撕取常山叶下表皮水装片，镜检，可见气孔周围有两个副卫细胞，其长轴与保卫细胞和气孔的长轴平行（附图 2-3）。

（2）直轴式气孔 撕取薄荷叶下表皮水装片，镜检，可见气孔周围的两个副卫细胞的长轴与保卫细胞和气孔的长轴相垂直（附图 2-4）。

（3）不等式气孔 撕取菘蓝叶或曼陀罗叶的下表皮水装片，镜检，可见气孔周围有 3 个副卫细胞，其中一个显著地比其他的两个细胞小（附图 2-5）。

（4）不定式气孔 撕取天竺葵叶的下表皮水装片，可见副卫细胞的数目不定，其形状与一般表皮细胞相似（附图2-6）。

4. 木栓 取甘草横切片观察，可见被染为棕红色，由排列紧密的扁平状木栓细胞构成的木栓层。注意在其横切片上，能否区分出栓内层和木栓形成层。

5. 皮孔 观察接骨木茎横切制片，注意在横切片周围的木栓层部分，可见有裂缝状缺口即皮孔，在皮孔内方有圆形或椭圆形、排列疏松的大型细胞，即补（填）充细胞。

三、机械组织

1. 厚角组织 取薄荷茎或南瓜茎横切片，镜检，注意在茎的棱角处的表皮下方，有数层细胞，其细胞只在角隅处增厚，增厚部分颜色较暗，并因相邻细胞数目不同而呈三角形或多边形，即厚角组织。

2. 厚壁组织

（1）纤维 取白藓皮粉末少许，用水合氯醛透化后，制成甘油装片，镜检，可见较多散在的纤维，呈长梭形，两头尖锐，有碎断，边缘光滑，壁极厚，木化，层纹极细密，胞腔线形，孔沟多不明显（附图2-7）。

（2）石细胞 取石榴皮粉末少许，按上述方法制成临时装片，镜检，可见类圆形或类方形的石细胞，壁厚，其胞腔层纹细密，孔沟较明显。

分别取白藓皮、石榴皮粉末少许，置载玻片上，滴加间苯三酚试液和盐酸，加盖玻片后镜检，注意其中纤维和石细胞变为何种颜色。

【实验报告】

1. 绘常山叶、薄荷叶、菘蓝叶、曼陀罗叶的气孔轴式图，注明各部位名称。
2. 绘曼陀罗叶或天竺葵叶、薄荷叶或菘蓝叶的非腺毛图。
3. 绘曼陀罗叶、薄荷叶的腺毛图。
4. 绘石榴皮的石细胞图，白藓皮的纤维图，注明各部位名称。

【思考题】

1. 腺毛和非腺毛有何区别？腺鳞具何形态特征？
2. 厚角组织和厚壁组织在形态和结构上有何异同？
3. 表皮细胞的特征是什么？气孔有何作用？

Experiment 2 Merismatic Tissue, Protective Tissue and Mechanical Tissue

Aim and demand

1. Grasp the unique qualities of the Merismatic tissue.
2. Understand the distribution and qualities of the Protective tissue.
3. Understand the structural and cellular qualities of the Mechanical tissue.

Experimental materials

Leaves of Changshan (*Dichroa febrifuga* Lour.), Bohe (*Mentha haplocalyx* Briq.), Xiyanghong (*Salvia splendens* Ker. Gawl.), Mantuoluo (*Datura metal* L.), Tianzhukui (*Pelargonium ortorum* Bailey), Songlan (*Isatis indigotica* Fort.) and Duxingcai (*Lepidium apetalum* Willd.). Powders of root skin of Baixian (*Dictamnus dasycarpus* Turcz.) and fruit peel of Shiliu (*Punica granatum* L.). Slide of vetical section of root tip of Yangcong (*Allium cepa* L.). Slide of cross section of Bohe (*Mentha haplocalyx* Briq.) stem, stem of Nangua (*Cucurbita moschata* (Duch.) Poiret), root of Gancao (*Glycyrrhiza uralensis* Fisch.) and elder stem of Jiegumu (*Sambucus willamsii* Hance).

Instruments and aplliances

Microscope, dissector, spirit lamp, chloral hydrate reagent, thin glycerine, trihydroxy - phenole reagent, hydrochloric acid.

Content and procedures

1. Merismatic tissue

Apicalmeristems: Observe slide of vertical section of root tip of *Allium cepa*, firstly to observe under low powered microscope and to find the smallest cell and darkest cylindrical growth cone, that is the part of root tip meristems. The root tip meristems can be roughly divided into two parts: those at the top of the tip is a small group of the smaller and youngest cells, which can divide strongly and permanently, named promeristem, it has the thin cell wall, concentrated cytoplasm and big nucleus, vacuoles are small and numerous, the cells are in organized, without intercellular space. The second part of cells follow the promeristem are beginning to specialize, the surface of it is pro - epidermis, the central column part is procambium, the part between the pro - epidermis and procambium is ground meristem. Ahead the growth cone there is the root cap.

2. Protective tissue

2.1 Epidermis Tear a small piece from back epidermis of a *Mentha haplocalyx* leaf, make into a temporary slide. Look under microscope, you can see that the epidermis is made up of a thin layer of flat parenchyma cell, the cells are packed tightly, they do not contain chlorophyll. Look to the surface of the cell you can see that the side wall of the cell is wave shaped, the cells embed from each other, joined firmly. At the same time, pay attention to whether you can see cuticle, trichome and stoma in the epidermis or not.

2.2 Trichome

2.2.1 Non - glandular trichome Tear a little piece of back epidermis of leaves of *Isatis indigotica* or *Pelargonium ortorum*, and make a temporary water slide respectively, then to observe under micoroscope. You can see many non - glandular trichomes with sharp narrow top, and count the numbers of the cell that make up this structure (Figure 2 - 1).

2.2.2 Glandular trichome

2.2.2.1 Observe the water slide of back epidermis of leaves of *Mentha haplocalyx* or *Salvia splendens*, notice that there are two types of glandular trichome, one type has 1 ~ 2 cells of secretory cell and 1 ~ 2 cells of stalk (Figure 2 - 1). Another type is named glandular scale, it's secretory cell is consist of 6 ~ 8 cells, little flat spheroidal shape, arrange in a same plane. Around the secretory cell, there are cuticles (Figure 2 – 2). Volatile oil store between the cuticles and secretory cell, the stalk is very short.

2.2.2.2 Observe water slide of back epidermis of *Datura metal* leaf, notice the appearance and structure of glandular trichome.

2.3 Stoma

2.3.1 Paracytic type Make water slide of back epidermis of *Dichroa febrifuga* leaf, to observe under microscope, two subsidiary cell should be visible around the stoma, its long axis is parallel to the long axis of guard cell and stoma (Figure 2 – 3).

2.3.2 Diacytic type Make water slide of back epidermis of *Mentha haplocaly* leaf, to observe under microscope, two subsidiary cell should be visible around the stoma, its long axis is vertical to the long axis of guard cell and stoma (Figure 2 – 4).

2.3.3 Anisocytic type Make water slide of back epidermis of leaves of *Isatis indigotica* or *Datura metal*, to observe under microscope, three subsidiary cell should be visible around the stoma, one of which appears smaller than two others (Figure 2 – 5).

2.3.4 Anomocytic Type Make water slide of back epidermis of *Pelargonium ortorum* leaf, to observe under microscope, you can find that numbers of the subsidiary cells around the stoma are not regular, its shape is similar to the common epidermis cell (see figure 2 – 6).

2.4 Cork Take cross section of *Glycyrrhiza uralensis* root to observe, you can see phellem that dyed into brown red. It is made up of firmly – arranged flat cork cells. Pay attention to whether you can distinguish phelloderm and phellogen in cross section.

2.5 Lenticel Observe cross section of stem of *Sambucus willamsii*, in the outside part of the phellem, there are many crack shaped gap that is the lenticel. Inside the lenticel, there are many circular or ellipse loose arranged large cells, named stuff cell.

3. Mechanical tissue

3.1 Collenchyma Observe cross section of stem of *Mentha haplocalyx* or stem of *Cucurbita moschata* under microscope, notice that under the epidermis of stem arris, there are several layers of cells which thicken only in the corner. The thicken part is dark, and usually is triangle or polygon shape, that is collenchyma.

3.2 Sclerenchyma

3.2.1 Fiber Take a little amount of powder of root skin of *Dictamnus dasycarpus*, transparentized with chloral hydrate, then to make a glycerine slide. Observe under a micoroscope, there are many scatterly distributed fibers. The fibers are long shuttle shape, both ends are sharp, the edge is smooth, very thick wall is lignification, lamellae is close, cell cavity is line shape, pit canal numerous and not clear (Figure 2 – 7).

3.2.2 Sclereid Operate as above to make temporary slide of powder of fruit peel of *Punica granatum.* You can see near spherical or square sclereid with thick wall. Its lamellae of wall is close and pit canal is obvious (Figure 2 – 8).

Take little amount of powder of root skin of *Dictamnus dasycarpus* and fruit peel of *Punica granatum* respectively, put them together on a slide, then drop some phloroglucinol reagent and hydrochloric acid on the powder, cove a cover glass, then to observe. Pay attention to the colour change of fiber and sclereid.

Experiment report

1. Draw the stoma type of leaves of *Dichroa febrifuga*, *Mentha haplocalyx*, *Isatis indigotica* and *Datura metal*, note the name of each part.

2. Draw the non – glandular trichome of *Datura metal* or *Pelargonium ortorum*, *Mentha haplocalyx* or *Isatis indigotica*.

3. Draw theglandular of Datura metal and *Mentha haplocalyx*.

4. Draw sclereid of root skin of *Dictamnus dasycarpus* and fruit peel of *Punica granatum*.

Questions

1. How to distinguish glandular trichome and non – glandular trichome? List the morphological characteristics of gland scale.

2. How many differences of form and structure are there between collenchyma and sclerenchyma?

3. List characteristics of epidermis cell and the functions of stoma.

实验三　输导组织、分泌组织和维管束的构造与类型

【目的要求】

1. 掌握输导组织的细胞形态和结构特征，熟悉导管的各种类型及结构。
2. 掌握分泌组织的细胞形态和结构特征，了解其不同类型的分布位置。
3. 了解维管束的基本构造，识别各种维管束的类型。

【实验材料】

鲜姜、玉竹根茎、陈皮及其横切片、人参根横切片、蒲公英根纵切片、小茴香果实横切片、松茎横（纵）切片、向日葵茎横（纵）切片、南瓜茎横（纵）切片、东北贯众根茎横切片、石菖蒲根茎横切片、玉米茎横切片、毛茛根横切片、油松茎和葡萄茎的解离组织。

【仪器用品】

显微镜、解剖用具、稀甘油、墨汁、蒸馏水。

【内容与方法】

一、输导组织

1. 管胞　取松茎纵切片，在显微镜下观察，可见管胞呈长管状，两端常偏斜，两相邻管胞侧壁上的纹孔相同，为具缘纹孔，因具纹孔塞，故表面观呈3个同心圆（附图3-1）。

再观察松茎的解离组织甘油装片，管胞形态结构比纵切片更加清晰。

2. 导管

（1）在低倍镜下观察向日葵茎纵切片，选择被番红染成红色的、具有花纹而成串的管状细胞，即各种类型的导管。注意每个导管分子均以端壁打通后形成的穿孔相互连接、上下贯通。能否区分出环纹、螺纹、梯纹、网纹和孔纹导管。（附图3－2，附图3－3）

（2）取葡萄茎解离组织少许，制成临时甘油装片，注意其导管为何种类型。

3. 筛管及伴胞

（1）取南瓜茎横切片，先用手持放大镜粗略观察，可见在南瓜茎中央有一个五角星状的大空腔（髓腔），腔的周围有排列成环的10多个维管束，内轮五个较大，外轮五个较小，正对着南瓜茎横切面上突起的五个棱。然后换低倍镜对维管束详细观察：南瓜茎的维管束为双韧型，即每个维管束的中部为被番红染成红色的木质部，其中有几个明显的大导管，其内、外两侧为侧生分生组织，即维管形成层，它们两端为韧皮部，筛管和伴胞都分布在此处。移动载玻片，在外韧皮部中寻找筛管和伴胞，筛管为多边形薄壁细胞，常被固绿染成蓝绿色，在它旁边常贴生有一个小型的伴胞。个别筛管正好切在筛板处，可见染色较深的筛板和其上的筛孔（附图3－4）。

（2）观察南瓜茎的纵切制片，可见筛管是由许多长管状的细胞，即筛管分子连接而成，两细胞连接的端壁上有筛板及其上的筛孔。在筛管旁边呈梭形的细胞即伴胞。

二、分泌组织

1. 分泌细胞

（1）油细胞　取鲜姜作徒手切片，制成临时水装片，镜检，可见在薄壁细胞之间，杂有许多类圆形的油细胞，胞腔内含有淡黄绿色挥发油滴。

（2）黏液细胞　取玉竹根茎作徒手切片，置载玻片上，加墨汁一滴，盖上盖玻片，镜检，可见黏液细胞呈无色透明状，常含草酸钙针晶。其他细胞和内含物均显黑色。

2. 分泌腔　观察陈皮（橘的成熟果皮），可见外表有圆形或凹入的小点即分泌腔，因内贮挥发油而称为油室。再观察陈皮的横切制片，可见在果皮近中部有大型腔穴，即油室，在腔穴周围可见到有部分破裂的分泌细胞（附图3－5）。

3. 分泌道

（1）观察小茴香果实横切制片，可见果实由两个分果组成。分果横切面略呈五边形，具5棱，每两棱间有一个小圆腔，即油管，每棱脊内有1个维管束，在两分果接合面有2个油管（附图3－6）。

再观察人参根的横切片，注意其树脂道的形状及分布情况。

（2）观察松茎横切制片，可见在被番红染成红色的木质部中，有许多整齐排列的分泌细胞围成的大圆腔，即分泌道，因其内贮藏树脂而称树脂道（附图3－7）。

4. 乳汁管　观察蒲公英根纵切片，在显微镜下可见染成颜色略深的分枝状的乳汁管。

三、维管束的构造及类型

取向日葵茎横切片，研究它的环列维管束，观察其中一束。

首先找出韧皮部的特征——筛管及伴胞，木质部的特征——导管，注意二者的排列方式；然后观察木质部和韧皮部之间是否具有细胞为扁长方形、排列紧密的维管形成层，指出这种维管束属于哪种类型。

同样方法观察下列材料。

（1）玉米茎横切片：有限外韧式维管束。

（2）南瓜茎横切片：双韧式维管束。

（3）东北贯众根茎横切片：周韧式维管束。

（4）石菖蒲根茎横切片：周木式维管束。

（5）毛茛根横切片：辐射式维管束。

【实验报告】

1. 绘导管、管胞及筛管图，注明材料名称。
2. 绘人参根茎横切片中的树脂道图。
3. 绘各种维管束类型简图，注明各部位名称及材料名称。

【思考题】

1. 管胞、纤维及导管在形态结构上有何异同？
2. 什么叫溶生式分泌腔和裂生式分泌腔？
3. 维管束一般构造如何？可分为几种类型？

Experiment 3　Conducting Tissue, Secretory Tissue, Structure and Types of Vascular Bundle

Aim and demand

1. Grasp the cellular shape and structural features of conducting tissue, familiarize yourself with the types and structure of vessel.

2. Grasp the cellular shape and structural features of secretory tissue; familiarize yourself with the distributed position of different types.

3. Find out the basic structure of vascular bundle, be able to distinguish the different types of vascular bundle.

Experimental materials

Freshrhizome of Jiang (*Zingiber offcinales* Rosc.), Yuzhu (*Polygonutum odoratum* (Mill.) Druce), fruit peel and it's cross section of Ju (*Cittrus reticulata* Blanco), slide of cross section of Renshen (*Panax ginseng* C. A. Meyer) root, vertical section of root of Pugongying (*Taraxacum Mongolicum* Hand - Mazz.), cross section of fruit of Xiaohuixiang (*Foeniculum vulgare* Mill.), Slide of cross and vertical section of stem of Yousong (*Pinus tabulaeformis* Carr.), Xiangrikui (*Helianthus annuus* L.) and Nangua (*Cucurbita moschata* (Duch.) Poiret). Cross section of rhizome of Dongbeiguanzhong (*Dryopteris crassirhizoma* Nakai), rhizome of Shichangpu (*Acorus tatarinowii* Schott), stem of Yumi (*Zea mays* L.)

and root of Maogen (*Ranunculus japonicus* Thunb.) . Dissociation tissue of stem of Yousong (*Pinus tabulaeformis* Carr.) and Putao (*Vitis vinifera* L.) .

Instruments and aplliances

Microscope, dissecting tools, thin glycerine, ink, distilled water.

Contents and procedures

1. Conducting tissue

1. 1 Tracheid To observe a slide of vertical section of *Pinus tabulaeformis* stem under the microscope, long tubular shaped tracheid should be visible, the two ends are usually inclined, the pit on the walls of two adjacent tracheid are connected, this is bordered pit. Because it has pit plug, you can see three concentric circles through its surface (see Figure 3 – 1) . Then observe the glycerine slide of *P. tabulaeformis* stem dissociation tissue. The shape and structure of tracheid should be more clearer than vertical section.

1. 2 Vessel

1. 2. 1 Observe slide of vertical section of *Helianthus annuus* stem under low power magnification. Choose the tubular cells that contain decorative and have been dyed red, that are the vessels. Notice that each vessel element connected with each other through performations that punch the two end wall. At last, to distinguish the types of annular vessel, helical vassel, scalariform vessel, reticulate vessel and pitted vessel (Figure 3 –2, 3 –3) .

1. 2. 2 Take alittle amount of dissociation tissue of *Vitis vinifera* stem, make temporary glycerine slide, to observe the types of vessels.

1. 3 Sieve tube and companion cell

1. 3. 1 Take slide of cross section of *Cucurbita moschata* stem, firstly observe roughly with a magnifying glass. A star shaped large marrow cavity should be visible in the center of the pumpkin stem. About 10 vascular bundles form a ring around the cavity. The inner 5 are larger and the outer 5 are relatively smaller. Directly opposite to the cut surface of the pumpkin stem are 5 raised arris. Afterwards, switch to low powered microscope to carefully observe the vascular bundle. The type of pumpkin stem vascular bundle is bicollateral vascular bundle, in the middle part of the vascular, there is xylem which was dyed into red by safranine O, among the xylem there are some distinct large vessels. Inside and outside of it is the lateral meristems, that is the vascular cambium. The two ends of the vascular bundle are phloem, both sieve tube and companion cell are distribute here. Move the slide, to look for sieve tube and companion cell in the out phloem. Sieve tube are polygonal parenchyma cells, often dyed bluish green by Green – fixed dye solution. On its side, there usually grows a small companion cell. A few sieve tube happens to be cut at the sieve plate, and heavily dyed sieve plate and sieve pore on it can be seen (Figure 3 –4) .

1. 3. 2 Observe vertical section of *Cucurbita moschata* stem, it can be seen that sieve tube is made up of many long tubular cells connected by sieve – tube element. At the end wall at which the two cells connect are sieve plate and sieve pore on it. Near the sieve tube, there is a

shuttle shaped cell, this is the companion cell.

2. Secretory tissue

2.1 secretory cells

2.1.1 oil cell Take fresh rhizome of *Zingiber offcinales* and make free hand slide, make it into temporary water slide, then to observe under microscope, it can be seen that between parenchyma there are many near circle shape oil cell, its cell cavity contains light yellow - greenish volatile oil drop.

2.2.2 mucilage cell Take rhizome of *Polygonutum odoratum* to make a free hand slide, add one drop of ink on the sample, place cover glass on it, then to observe under microscope. You can see that mucilage cell are in a colorless clear state, often containing acicular crystal of calcium oxalate, other cells and contents appear black.

2.2 secretoy cavity Observe mature fruit peel of *Cittrus reticulata*, it can be seen that there are circle or concave little spot on the surface, that are secretory cavities. Because it contains volatile oil, named oil cavity. Then observe cross section of tangenire peel, you can see that there are large cavities near the center of the peel, that are oil cavities. There are also partly split secretory cells around the cavity (Figure 3-5).

2.3 secretory duct

2.3.1 Observe cross section of *Foeniculum vulgare* fruit, you can see that fruit are made up of two parts. The cross section of the schizocarp is pentagonal in shape, and has 5 arrises. Between each arris, there is a small circle cavity, that is the oil duct. In each ridge of the arris, there is a vascular bundle. In the contact surface of the two schizocarps, there are two oil ducts (Figure 3-6).

Observe cross section of *Panax ginseng* root, to pay attention to the shape and distribution of the secretory duct.

2.3.2 Observe the cross section of *Pinus tabulaeformis* stem, you can see that among the xylem that dyed into red by safranine O there are many large circle cavities which were surrounded by secretory cells. The secretory cells arranges in good order. This cavity is secretory duct named resin duct for it contains resin (Figure 3-7).

2.4 Laticifer Observe *Taraxacum Mongolicum* root vertical slice under microscope, branched and darkly dyed laticifers are visible.

3. Structureand types of vascular bundle

Take *Helianthus annuus* stem cross slice, to look for its annular vascular bundles, then to observe one of the bundle. Firstly to look for the distinguishing features of the phloem: sieve tube and companion cell; xylem's distinguishing features: vessel. Pay attention to the order of the two. Then observe whether or not there are flat, rectangular and tightly packed vascular cambium cells between phloem and xylem, point out which type of vascular bundle this belongs to.

Observe slices below use the same method as above:

Cross section of *Zea mays* stem：closed collateral vascular bundle.

Cross section of *Cucurbita moschata* stem：bicollateral vascular bundle .

Cross section of male *Dryopteris crassirhizoma* rhizome：amphicribrial vascular bundle.

Cross section of *Acorus tatarinowii* rhizome：amphivasal vascular bundle.

Cross section of *Ranunculus japonicus* root：radial vascular bundle.

Experiment report

1. Draw vessels，tracheid and sieve tube，note names of materials.

2. Draw a resin duct of *Panax ginseng* root or vertical section of *Pinus tabulaeformis* stem.

3. Draw sketch of all types vascular bundles you have observed，note names of different parts and the names of materials.

Questions

1. What is the difference in structure among tracheid，fiber and vessel?

2. What are dissolve－formated secretory cavity and separate－formated secretory cavity?

3. What is the normal structure of vascular bundle? How many types are there?

实验四　根的形态、初生构造、次生构造与异常构造

【目的要求】

1. 掌握识别根的外部形态和类型及变态根的形态、特征和类型。
2. 了解根尖的解剖构造及分区。
3. 掌握双子叶植物根和单子叶植物根的初生构造及异同点。
4. 掌握双子叶植物根的次生构造特征。
5. 了解根的异常构造（三生构造）。

【实验材料】

桔梗、蒲公英、高山积雪、大豆、薏苡、葱、徐长卿、芍药、板蓝根、白芷、何首乌、天门冬、萱草、龟背竹、吊兰等植物的根；菟丝子、桑寄生或槲寄生带寄主的标本、络石藤的标本；玉米或洋葱根尖的纵切面的永久切片；毛茛、辽细辛、鸢尾、百部、芍药、人参、牛膝、黄芩根及何首乌块根的横切片。

【仪器用品】

镊子、放大镜、解剖镜、显微镜。

【内容与方法】

一、根的形态特征和类型

1. 直根系　观察桔梗、蒲公英、高山积雪、大豆的根系，注意分辨主根、侧根和纤维根。

2. 须根系　观察薏苡、葱、徐长卿的根系，注意是否有主根，有无不定根的发生，这种根系是如何形成的。

二、根的变态

1. 贮藏根 观察桔梗、芍药、板蓝根、欧白芷等植物的根，可见其直根已膨大成不同形状，这种变态根主要适应于贮藏大量的营养物质。

2. 块根 观察何首乌、天门冬、萱草等植物的根，注意与上一种变态根的区别，指出其变态块根的不同来源。

3. 气生根 观察龟背竹、吊兰露在空气中的不定根；观察玉米、薏苡渗入土壤中的不定根，指出它们生理功能有何不同。

4. 寄生根 观察带有寄主的菟丝子、桑寄生或槲寄生标本，注意它们的根均伸入寄主的茎内，其中菟丝子不含叶绿体，不能制造养料。

5. 攀援根 观察络石藤，注意由茎上产生能攀附他物的不定根。

三、根尖的构造和分区

取玉米根尖纵切面的永久切片，在显微镜下观察。由根尖逐渐向上辨认以下各区，并注意其细胞特点。

1. 根冠 在根尖的最先端，略呈三角形，套在生长点之上。是一群薄壁细胞，排列不整齐，形小而质浓，能为根冠不断形成新细胞。

2. 分生区 在根冠之内，长仅1～2mm，由排列紧密的小型多面体细胞组成。细胞壁薄，核大，质浓，分裂能力很强，属分生组织。

3. 伸长区 位于分生区上方，到出现根毛的地方，一般长2～5mm左右，细胞中液泡大量出现，除沿长轴方向迅速伸长外，同时逐步分化成不同的组织，向成熟区过渡。

4. 根毛区（成熟区） 在伸长区的上方，细胞分化成熟，并形成了各种初生组织，表面密生根毛。

四、根的初生构造

1. 双子叶植物根的初生构造 取毛茛根的横切片，在低倍镜下区分出表皮、皮层和维管柱三大部分，然后转换高倍镜由外向里仔细观察。

（1）表皮 为最外一层薄壁细胞，排列整齐紧密，没有细胞间隙。注意是否有角质层及根毛。

（2）皮层 在表皮以内，占根的大部分，分为3层。

外皮层：紧靠表皮下方，排列较紧密，略呈切向延长的1～2层细胞。

中皮层：介于外皮层和内皮层之间，由排列疏松的多层薄壁细胞组成。

内皮层：皮层的最内一层，细胞排列紧密，可见到被染成红色的凯氏点以及对着木质部束的通道细胞。

（3）维管柱 为内皮层以内的所有组织，细胞较小而密集，由三个部分组成（附图4－1）。

中柱鞘：由紧贴内皮层里的1～2层薄壁细胞组成，排列整齐而紧密。侧根、木栓形成层和维管形成层的一部分均发生于中柱鞘。

初生木质部：呈四束，即四原型。其导管被染成红色，每束导管口径大小不一，靠近中柱鞘的导管口径小，称原生木质部；近根中心的导管口径大，称后生木质部，这是根的初生构造特征之一。

初生韧皮部：位于两初生木质部之间，与初生木质部相间排列成辐射状，构成辐射维管束，这也是根的初生构造特征之一。初生韧皮部被染成绿色，细胞多角形。

同样观察细辛根横切片，注意区分表皮、皮层和维管柱。其初生木质部通常三原型。

2. 单子叶植物根的初生构造 观察百部根的横切片，由外向里观察。

（1）根被 有3~4列细胞组成，细胞壁具条纹状木栓化纹理。

（2）皮层 宽广，由薄壁细胞组成。内皮层明显，可见凯氏带。

（3）维管柱 位于中央，占根的小部分，包括中柱鞘、初生木质部、初生韧皮部和髓。

中柱鞘：紧贴内皮层，由1~2列小型薄壁细胞组成。

初生木质部束和初生韧皮部束：各19~27个，相间排列成辐射维管束。韧皮部束内侧有单个或2~3个成束的非木化纤维；木质部导管类多角形，偶有单个或2~3个并列的导管分布于髓部外缘，作二轮列状。

髓：位于维管柱中心，散有单个或2~3个成束的细小纤维。

同样方法观察鸢尾根横切片，并比较二者有何区别。

五、双子叶植物根的次生构造

取芍药根的横切片，从外向里逐层观察（附图4-2）。

1. 周皮 为最外方的数层细胞，由木栓层、木栓形成层和栓内层组成。

（1）木栓层 由数列排列整齐、扁长方形的木栓细胞组成，是没有细胞核的死细胞，常呈浅棕色。

（2）木栓形成层 是由中柱鞘细胞恢复分生能力而产生的，在切片中不易分辨。

（3）栓内层 狭窄，为2~3列呈切向延长的大型薄壁细胞，属于基本组织。初生韧皮部已被挤坏，分辨不清。

2. 次生维管组织 为形成层活动产生的组织。

（1）次生韧皮部 为周皮以内，被染成绿色的部分，甚宽。包括筛管、伴胞和韧皮薄壁细胞，其中夹杂有少量略呈红色的韧皮纤维。注意在横切面上韧皮薄壁细胞与筛管形态相似，常不易区分。韧皮射线呈放射状的倒三角形，是由许多韧皮薄壁细胞在径向上排列形成。

（2）形成层 在次生韧皮部内方，由数列排列紧密、整齐的扁长方形薄壁细胞组成。形成层只有一列细胞，但由于它向内、向外分裂迅速，刚产生不久的衍生细胞，尚未分化成熟，故在横切面上看到的是多列细胞组成的“形成层区”。

（3）次生木质部 在形成层以内，包括导管、管胞、木薄壁细胞和木纤维。在横切面上，导管最容易辨认，它们是一些被染成红色的、口径大小不一的类圆形或多边形的死细胞，放射状排列。木射线由 1～2 列薄壁细胞组成，在木质部中也呈放射状排列，并与韧皮射线相连接，组成维管射线，起横向运输作用。

在次生木质部的内方，根的中心部位，为初生木质部。其导管口径细小，呈类圆形。

同样观察人参根横切片，其构造与芍药相似，其特点是次生韧皮部有树脂道。木质部导管多呈单列、径向稀疏排列。栓内层、木薄壁细胞和木射线中有草酸钙簇晶。

六、根的异常构造

（1）观察何首乌块根，在其呈淡黄棕色或淡红棕色的横断面上，可见皮部有 4～11 个类圆形的异型维管束环列，形成“云锦花纹”。然后在显微镜下观察何首乌块根的横切制片，从外向里区分出周皮、薄壁细胞组织、排成一圈大小不等的圆环异型维管束和中央正常维管束。形成层呈环状，异型维管束多为复合型，少数为单个维管束，均为外韧式（也有文献认为是周韧式）。根的中央为大型正常维管束，亦为外韧型，中心部分为初生木质部。

（2）观察怀牛膝根，可见其横断面平坦，角质样，淡黄色，木质黄白色，其外有众多小点，排成 2～4 轮，即异型维管束。然后在显微镜下观察其横切制片，其最外为木栓层，由 4～8 列扁平的木栓化细胞组成，木栓层以内，为数层薄壁细胞。维管组织占根的大部分，分布有多数异型维管束，断续排成 2～4 轮，最外轮的异型维管束较小，异常形成层几乎连接成环，内轮的异型维管束较大，均为外韧型。根中央为正常维管束，初生木质部常为二原型（附图 4－3）。

（3）观察黄芩老根横切片，在中央木质部中有木栓化细胞环。

【实验报告】

1. 列表记录观察到的变态根类型。
2. 绘毛茛根、百部根的横切片简图，并注明各部分名称。
3. 绘芍药根的横切片简图，注明各部分名称。
4. 绘牛膝或何首乌根的横切片简图，注明各部分名称。

【思考题】

1. 植物的根为什么会发生变态？这种变态有何意义？
2. 双子叶植物和单子叶植物的初生构造有什么异同点？
3. 根的初生构造和次生构造有何区别？次生构造是如何形成的？
4. 根的异常构造是如何形成的？

Experiment 4　Formation, Primary Structure, Secondary Structure and Anomalous Structure of Root

Aim and demand

1. Grasp the morphological characters and types of root and root modifications.
2. Understand the dissect structure and zone of root tip.

3. Grasp the primary structure and difference of root of dicotyledon and monocotyledon.

4. Grasp the secondary structure of dicotyledon.

5. Understand anomalous structure of root (tertiary structure) .

Experiment materials

The materials of root: Roots of Jiegeng (*Platycodon grandiflorum* (Jacq.) A. DC.), Pugongying (*Taraxacum monglicum* Hand. – Mazz.), Gaoshanjixue (*Euphorbia pekinensis* Rupr.), Dadou (*Glycine max* (L.) Merr.), Yiyi (*Coix lacryma – jobi* L. var. ma – yuen (Roman. Stapf), Cong (*Allium fistulosum* L.), Xuchangqing (*Cynanchum paniculatum* (Bunge) Kitag.), Shaoyao (*Paeonia lactiflora* Pall.), Banlangen (*Isatis indigotica* Fort.), Baizhi (*Angelica dahurica* (Fish. ex Hoffm.) Benth et Hook. f.), Heshouwu (*Polygonum multiflorum* Thunb.), Tianmendong (*Asparagus cochinchinensis* (Lour.) Merr.), Xuancao (*Hemerocallis fulva* (L.) L.), Guibeizhu (*Monstera deliciosa* Liebm.), Diaolan (*Chlorophytum capense* (L.) Voss.); samples with host of a parasite of Tusizi (*Cuscuta chinensis* Lam.), Sangjisheng (*Taxillus chinensis* (DC.) Danser) or Hujisheng (*Viscum coloratum* (Kom.) Nakai); sample of Luoshiteng (*Trachelospermum jasminoides* (Lindl.) Lem.); vertical section of root tip of Yumi (*Zea mays* L.) or Yangcong (*Allium cepa* L.); cross section of root of Maogen (*Ranunculus japonicus* Thunb.), Liaoxixin (*Asarum heterotropoides* Fr. Schmidt var. mandshuricum (Maxim.) Kitag.), Yuanwei (*Iris tectorum* Maxim.), Baibu (*Stemona sessilifolia* Miq. Franch. Et Sav.) . Shaoyao (*Paeonia lactiflora* Pall.) and Renshen (*Panax ginseng* C. A. Mey.); cross section of Niuxi (*Achyranthes bidentata* L.), Huangqin (*Scutellaria baicalensis* Georgi); tuber of Heshouwu (*Polygonum multiflorum* Thunb.).

Instruments and aplliances

Tweezers, magnifying glass, anatomical lens, microscope.

Contents and procedures

1. The morphological characters and types of root

1. 1 Tap root system Observing the roots of *Platycodon grandiflorum*, *Taraxacum monglicum*, *Euphorbia pekinensis* and *Glycine max*, to distinguish main root, lateral root and fibrous root.

1. 2 Fibrous root system Observing the roots system of *Coix lacryma – jobi*, *Allium fistulosum*, *Cynanchum paniculatum*, and paying attention to observe main root and adventitious root or not, how to consist of the root system.

2. Root modifications

2. 1 Storage root Observing the roots of *Platycodon grandiflorum*, *Paeonia lactiflora*, *Isatis indigotica*, *Angelica dahurica*, and may see main root expend to different forms. These abnormal roots are fit to store up lots of vegetative material.

2. 2 Root tuber Observing the roots of tuber of *Polygonum multiflorum*, *Aspaargus cochinchinensis*, *Hemerocallis fulva* and comparing with upper root, to point out different origin of abnormal root tuber.

2. 3 Aerial root Observing adventitious roots of exposed to air of *Monstera diliciosa*, *Chlorophytum capense*, and observing adventitious roots of extending to earth of *Zea mays* and *Coix lacryma -jobi*, to piont out what different in their physiological function.

2. 4 Parasitic root Observing sample with host of a parasite of *Cuscuta chinensis*, stem of *Taxillus chinensis*, *Viscum coloratum*, but *Cuscuta chinensis* hasn't chloroplast, and can't produce nutriment.

2. 5 Climbing root Observing *Trachelospermum jasminoides* to pay attention to climbing adventitious root come from stem.

3. The dissect structure and zone of root tip

Observing radial section of root tip of Zea mays, to distinguish every zone from root tip, and paying attention to observe the character of cell.

3. 1 Root cap It is a triangle part on the tip of root tip , and is a lot of packed regularly cells. Cells are small and cytoplasm is thick, and can produce new cells for root cap.

3. 2 Division zone It consists of small closely polyhedron of cells in the root cap , its length is only 1 - 2mm. Cell wall is thin, and cytoplasm is thick. Their division ability is strong and belongs to meristem.

3. 3 Elongation zone It lies on the division zone , extend to the place of appeared root hair. Its length is 2 - 5mm. There are lots of vacuoles in the cell, they extend quickly on the major axis, and divide to different tissues, and transition to maturation zone.

3. 4 Maturation zone It lies on the elongation zone. The cells divided to maturation, and form to all kinds of primary tissues, Its surfaces are covered lots of root hair.

4. Primary structure of root

4. 1 Primary structure of dicotyledon root Bringing cross section of root of *Ranunculus japonicus.* to distinguish epidermis, cortex and vascular cylinder under the lower microscope, then turn to the higher microscope to observe from outer to inter.

4. 1. 1 Epidermis It is a single layer of densely packed cells in the outer, without intercellular spaces , and observe to have cuticle and root hair or not.

4. 1. 2 Cortex It lies in the epidermis, and occupies main spaces. It is divided into 3 parts.

Exodermis: It lies closely under the epidermis. 1 - 2 layer of cells are arranged closely, and extended to cutting.

Mesodermis: It lies between the cortex and endodermis, and consists of lots of loose layers and slice of cells.

Endodermis: It lies inside the cortex. Cells are arranged densely, and have red casparian and passage cell at the opposite xylem.

4. 1. 3 Vascular cylinder The whole tissues lie in the endodermis, cells are small and dense, and divided into 3 parts (Figure 4 - 1):

Pericycle: It consists of 1 - 2 layers of slice cells in the densely endodermis. The cells are arranged closely and regularly. Lateral root, phellogen and a part of vascular cambium origin vascular cylinder.

Primary xylem: It is made of 4 bundles, namelytetrach. Its vessels are red, and big or small of the mouth of every bundle vessel is different. The small mouth of vessel near pericycle is named protoxylem; the big mouth of vessel near center is named metaxylem. These are one character of primary structure.

Primary phloem: It lies between two primary xylem, and arrange to radial form with primary xylem, and compose to radial bundle. These are one character of primary structure. Primary phloem are green, and the cells are poly - angle.

Observing cross section of *Asarum heterotropoides* var. mandshuricum with the same method, and paying attention to distinguish epidermis cortex and vascular cylinder. Primary xylem is triarch.

4. 2 Primary structure of monocotyledon root Bringing slide of cross section of root of *Stemona sessilifolia.* from outer to inter to observe.

4. 2. 1 Quilt of root It consists of 3 - 4 lines of cells, and cells wall are striped suberizatal decorative pattern.

4. 2. 2 Cortex It is wide, and composed of slice cells. Endodermis and casparian dots are obvious.

4. 2. 3 Vascular cylinder It lies in the center, and occupies a little part of root, and includes the pericycle, primary xylem, primary phloem bundle and pith.

Pericycle: It consists of 1 - 2 layers of slice cells in the dense endodermis.

Primary xylem andprimary phloem bundle: They compose of 19 ~ 27 and arranged in radial bundle. There are 2 - 3 bundles fibers of unlignification in primary phloem bundle; The vessel of primary xylem are poly - angle, and there is single or 2 - 3 vessels of stand side by side out of pith in time, and become form of two lines.

Pith: It lies in the center of pericycle, and there is a thin fiber of single or 2 - 3 bundles.

Observingcross section of root of *Iris tectorum* by same mean, and compare in differences with each other.

5. The secondary structure of dicotyledon

Bringing cross section of root of *Paeonia lactiflora* to observe from outer to inter (Figure 4 - 2).

5. 1 Periderm It is composed of lots of layer cells at the outer, including phellem, phellogen and phellodem.

5. 1. 1 Phellem It consists of the suberification cells of regularly arranged, and cells are light brown rectangle and haven't nucleus.

5. 1. 2 Phellogen It comes into pericycle cells of storing divided ability, and isn't easy to distinguish in the section.

5. 1. 3 Phelloderm It is narrow, and composed of big and slice cells extended on the major axis, and belongs to ground tissue. Primary phloem is damaged and not easy to distinguish.

5. 2 Secondary vascular tissue It comes into the tissues of cambium.

5. 2. 1 Seconsary phloem It is dyed green and wide in the periderm, and includes sieve tube, companion cell and slice cells of phloem. There is a little red fiber of pholem in it. Slice cells of phloem and sieve tube are similar on the cross section, and not easy to distinguish. Pholem ray is radiative inverse triangle. It composes of lots of slice cells arranged on the radial axis.

5. 2. 2 Cambium It consists of flat rectangle cells of arranged regularly and closely in the phloem. Cambium is only a line cell, but it divides quickly to inner and outer, beared cell just aren't mature, so you can see "cambium zone" composed of lines of cells on the cross section.

5. 2. 3 Secondary xylem It includes the vessel, tracheid, slice cells of xylem and fiber of xylem in the cambium. Vessels are the easiest to distinguish on the cross section, they are red and similar round or poly - angle dead cells. Their mouths are big or small and their arrangement is radiative. Xylem ray consists of 1 - 2 lines slice cells, and they are radiative too. They relate to phloem ray and form vascular ray, have conducting ability on the crosswise.

Primary xylem lies in the secondary xylem and centre of root. Their vessels are similar round and their mouths are small.

Observing cross section of root of *Panax ginseng*, its structure is similar with *Paeonia lactiflora*. Its factor is that seconsary phloem have resin canal, and vessel of xylem are a line and radial loose arrangement, and have calcium oxalate cluster crystal in the phelloderm, slice cells of xylem and xylem ray.

6. Anomalous structure of root

6. 1 Observing root tuber of *Polygonum multiflorum*. It have 4 - 11 similar round anomalous vascular round in the cortex on the thin yellow - brown or red - brown cross section, and name "a cloud decorative". To distinguish periderm, slice cells, arranged a round big or small round anomalous vascular and normal vascular in the centre from outer to inter in the microscope. Cambium is round, anomalous vascular is complex, and a little is single vascular. They are a collateral bundle (or amphicribral bundle). The big normal vascular lies on the centre of root, and collateral bundle too. Primary xylem lies on the centre.

6. 2 Observing root tuber of *Achyranthes bidentata*. Has a lot of dots arranged 2 - 4 round, namely anomalous vascular, Its cross section is thin yellow flat and cuticle. The phellem lies at the outer on the cross section, and composed of 4 - 8 lines suberification cells. There are lots of stratas slice cells in it. Vascular tissues occupy a lot of spaces of root, and distribute a number of anomalous vascular, arranged 2 - 4 lines intermittently. The outer is smaller and anomalous cambium line to round. The inner is bigger and collateral bundle. Normal vascular lie on the centre of root, and primary xylem is diarch (Figure 4 - 3).

6.3 Observing the cross section of old root of *Scutellaria baicalensis* to find suberification cell round in the centre of xylem.

Experiment report

1. List the type of root modifications.

2. Draw the dissect structure of root of Ranunculus japonicus and *Stemona sessilifolia* and note the name of every section.

3. Draw the dissect structure of *Paeonia lactiflora*, and note the name of every section.

4. Draw the dissect structure of *Achyranthes bidentata* or *Polygonum multiflorum*, and note the name of every section.

Questions

1. Why do plant happen root modifications ? What is the meaning of the root modifications?

2. What are the differences of the primary structure of dicotyledon root and monocotyledon root?

3. What are the differences of the primary and secondary structure root? How is the secondary structure formed?

4. How is the anomalous structure of the root formed?

实验五　茎的形态和变态类型，茎的初生构造，双子叶植物和裸子植物茎的次生构造

【目的要求】

1. 掌握茎的外部形态和类型及变态茎的形态、特征和类型。
2. 掌握双子叶植物茎的初生构造。
3. 掌握双子叶植物木质茎和裸子植物茎的次生构造特征。
4. 掌握双子叶植物草质茎次生构造。

【实验材料】

茎的实物：紫丁香、胡桃楸、银杏、玉竹、白茅、姜、马铃薯、掌状半夏、东北天南星、洋葱、百合、唐菖蒲、假叶树、竹节蓼、天门冬、皂荚、丝瓜等植物的茎；向日葵茎、椴树茎的横切片；松树茎的横切、径向纵切、切向纵切片；薄荷和益母草茎的横切片。

【仪器用品】

镊子、放大镜、显微镜。

【内容与方法】

一、茎的形态特征

取2年生或3年生的木本植物紫丁香、胡桃楸、银杏等植物的枝条，观察其形态

特征。注意茎与根的区别。

二、茎的变态

1. 地下变态茎

（1）根状茎　观察玉竹、白茅、姜的根状茎，注意其外形，是否能分辨出节和节间，顶芽、侧芽及茎痕，鳞片状退化的叶。

（2）块茎　取马铃薯、掌状半夏、东北天南星等植物的块茎进行观察，分辨其顶芽及节上的芽眼，注意其缩短的节间。

（3）鳞茎　取洋葱、百合的鳞茎，首先将洋葱鳞茎纵切，观察其节间缩短的鳞茎盘，注意其上的肉质及膜质鳞片叶、顶芽及腋芽。观察百合的鳞茎，注意较肥厚的肉质鳞叶。

（4）球茎　观察唐菖蒲的球茎，为球形或扁球形的肉质地下茎。是由地下侧枝末端膨大而形成。

2. 地上变态茎

（1）叶状茎　观察假叶树、竹节蓼、天门冬等植物，注意茎扁平变态为绿色的叶状体，叶已退化成鳞片状。

（2）刺状茎　观察皂荚的刺，注意其位置生于叶腋，由腋芽发育形成，不易剥落。

（3）茎卷须　观察丝瓜茎卷须标本，注意其发生位置。

三、茎的初生构造

取向日葵幼茎横切片，先在低倍镜下区分出表皮、皮层和维管束三部分；维管束环状排列为一圈，束间有髓射线，中央为宽大的髓。然后转换高倍镜逐层观察（附图5－1）。

1. 表皮　为一层排列整齐、紧密的扁长方形的薄壁细胞组成，其外壁角质加厚，有时可见非腺毛。

2. 皮层　为多层薄壁细胞，具细胞间隙。与根的初生构造相比，所占比例很小。靠近表皮的几层细胞较小，是厚角组织，细胞在角隅处加厚，细胞内可见被染成绿色类圆形叶绿体，其内为数层薄壁细胞，其中有小型分泌腔。皮层的最内一层细胞无凯氏带的分化。

3. 维管柱　所占面积宽广，包括维管束、髓射线和髓。

（1）维管束　为数个大小不等的无限外韧型维管束，排成一轮，每个维管束由初生韧皮部、束中形成层、初生木质部组成。

（2）初生韧皮部　位于维管束外方，其外侧有初生韧皮纤维，横切面呈多角形，壁明显加厚，但尚未木化，故被染成绿色。在初生韧皮纤维内方是筛管、伴胞和韧皮薄壁细胞。

（3）初生木质部　包括原生木质部和后生木质部，导管横切面类圆形或多角形，根据导管分子口径的大小和染色的深浅可以判断靠近茎中心的是原生木质部，导管口径小，发生早，染色深；而接近束中形成层的为后生木质部，导管口径大，发生较晚，染色浅淡。

（4）髓射线　是两个维管束之间的薄壁细胞，它外连皮层，内接髓部，具横向运输，兼有贮藏的功能。

（5）髓　位于茎的中央，也是维管柱中心的薄壁细胞，排列疏松，常具贮藏功能。

四、双子叶植物茎的次生构造

观察3~4年生椴树茎的横切片，由外向里可见（附图5-2）。

1. 周皮　明显，由木栓层、木栓形成层和栓内层组成。

（1）木栓形成层　是由皮层细胞恢复分裂能力以后形成的，细胞呈扁平状，细胞质浓。

（2）木栓层　在横切面上观察，是一些在同一半径线上排列整齐的扁平细胞。壁栓质加厚，是死细胞。

（3）栓内层　一般只有1~2层细胞组成，细胞质很浓，具细胞核，染色较重。

2. 皮层　较窄，仅有数层薄壁细胞组成，有时可观察到有些细胞内含有草酸钙簇晶。

3. 韧皮部　细胞排列成梯形（底部靠近形成层），与排列成喇叭形的髓射线薄壁细胞相间分布。在切片中，明显可见染成红色的韧皮纤维与被染成绿色的韧皮薄壁细胞、筛管和伴胞呈横条状相间排列。初生韧皮部已破坏。

4. 形成层　实为形成层区，呈环状，由4~5层排列整齐的扁长细胞组成。

5. 木质部　在形成层内方，在横切面上占有最大面积。主要是次生木质部。由于其细胞直径的大小和壁的厚薄不同，可以看到数轮同心轮层，即年轮。注意观察早材和晚材在组织构造上的区别。紧靠髓部周围的一群小型导管即初生木质部。

6. 髓　位于茎的中央，多由薄壁细胞组成，有的含草酸钙簇晶，有的含黏液和鞣酸，所以部分细胞染色较深。

7. 髓射线　由髓部薄壁细胞向外辐射状发出，直达皮层经木质部时，为1~2列细胞，至韧皮部时则扩大成喇叭状。

8. 维管射线　在每个维管束之内，由木质部和韧皮部中的横向运输的薄壁细胞组成，一般短于射线。位于木质部的称木射线，位于韧皮部的称韧皮射线。

五、双子叶植物草质茎的次生构造

取薄荷茎横切制片，可见茎呈四方形，在显微镜下由外到里仔细观察。

1. 表皮　为一层长方形表皮细胞组成，外被角质层，有的具毛（腺毛、非腺毛或腺鳞）。

2. 皮层　较窄，为数层排列疏松的薄壁细胞组成。在四个棱角内方，各有10余层厚角细胞组成的厚角组织，其细胞角隅处加厚明显，切片被染成绿色。内皮层明显，径向壁上可见被染成红色的凯氏点。

3. 维管柱

（1）维管束　由4个大的维管束（正对棱角）和其间较小维管束环状排列。韧皮部在外方，较窄，形成层成环，束间形成层明显。木质部在棱角处发达，导管单列，数行，纵向排列，在导管列之间为薄壁细胞组成的维管射线。

（2）髓　发达，由大型薄壁细胞组成。

（3）髓射线　为维管束间的薄壁细胞组成，宽窄不一。

此外，在茎的各部薄壁细胞中，有的还可见到扇形，具放射状纹理的橙皮苷结晶。

观察益母草茎横切片，其构造和薄荷茎相似，但皮层细胞含草酸钙小针晶。内皮层内侧有中柱鞘纤维散在。髓部细胞内含有草酸钙方晶和针晶。

六、裸子植物茎的次生构造

1. 观察3～4年生松树茎的横切片

（1）木质部　无导管，无典型的木纤维，主要有管胞、木薄壁细胞和木射线组成。

（2）韧皮部　无筛管和伴胞的分化，由筛胞、韧皮薄壁细胞和射线组成。

（3）在皮层和维管柱中，分布有许多树脂道。

2. 径向纵切面　注意年轮的排列方向、边材、心材、射线等。

3. 切向纵切面　对照1、2观察上述部分，注意有什么不同。

【实验报告】

1. 绘茎的外形图，并注明各部分名称。
2. 列表记录变态茎的类型。
3. 绘向日葵茎横切面简图，注明各部分名称。
4. 绘椴树茎、松树茎横片面简图，注明各部分名称。
5. 绘薄荷茎横片面简图，注明各部分名称。

【思考题】

1. 如何区分植物的根与植物的地下变态茎？
2. 双子叶植物茎的初生构造与根的初生构造有何不同？
3. 双子叶植物茎的次生构造是怎样形成的？
4. 双子叶植物茎的次生构造与裸子植物的次生构造有什么不同？

Experiment 5　Formation, Type and Modification of Stem; Primary Structure of Stem; Secondary Structure of Dicotyledon Stem and Gymnospermae Stem

Aim and demand

1. Grasp the outer formation and type of stem and formation, type, character of stem modifications.

2. Grasp the primary structure of stem.

3. Grasp the secondary structure of dicotyledon and gymnospermae.

4. Gain the secondary structure of herbaceous stem of dicotyledon

Experiment materials

The material of stem: Zidingxiang (*Syringa oblata* Lindl.), Hutaoqiu (*Juglans mandshurica* Maxim.), Yinxing (*Ginkgo biloba* L.), Yuzhu (*Polygonatum odoratum* (Mill.) Durce), Baimao (*Imperata cylindrical* Beauv. Var. *major* (Nees) C. E. Hubb.), Jiang (*Zingiber officinales Rosc.*), Malingshu (*Solanum tuberosum* L.), Banxia (*Pinellia ternata* (Thunb.) Breit), Dongbeitiannanxing (*Arisaema amurense* Maxim.), Yangcong (*Allium cepa* L.), Baihe (*Lilium brownii* F. E. Brown. var. *Viridulum* Backer), Tangchang pu (*Gladioulus gandavensis* Van Houtte), Jiayeshu (*Ruscus aculeatu* L.), Zhujieliao (*Homalocladium platycladium* (F. Muell.) Bailey), Tianmendong (*Asparagus cochinchinensis* (Lour.) Merr.), Zaojia (*Gleditsia sinensis* Lam.), Sigua (*Luffa cylindrica* (L.) Roem); cross section of stem of Xiangrikui (*Helianthus annuus* L.), Duanshu (*Tilia amurensis* Rupr.) transverse section, radial section and tangential section of Songshu (*Pinus tabulaeformis* Carr.), Bohe (Mentha haplocalyx Briq.) and Yimucao (*Leonurus heterophyllus* Sweet.).

Instruments and appliances

Tweezers, magnifying glass, microscope.

Contents and procedures

1. The morphological characters of stem

Observing 2 or 3 years stems of Syringa oblata, *Juglans mandshurica*, *Ginkgo biloba*, to distinguish the differences between stem and root.

2. Stem modifications

2.1 Underground stem modifications

2.1.1 Rhizome Observing the rhizomes of *Polygonatum odoratum*, *Imperata cylindrical*, *Zingiber officinales*, to find out different of node, internode, terminal bud, lateral, axillary bud and degenerative scale leaf.

2.1.2 Tuber Observing the tubers of *Solanum tuberosum*, *Pinellia ternata* And *Arisaema amurense*, and distinguishing their terminal bud, bud eye of node and shortened internode.

2.1.3 Bulb Bringing and cutting bulbs of *Allium cepa* and *Lilium brownii.* var. *Viridulum*, and observing shortened bulb plate, to pay attention to their succulent and membranaceous scale leaf, terminal bud and axillary bud. Observing the succulent tuber of lily.

2.1.4 Corm Observing the corm of *Gladioulus gandavensis.* It is round or flat - round succulent underground stem, and comes from end expended underground branch.

2.2 Modification of stem on the ground

2.2.1 Phylloclade Observing green leaf forms of *Ruscus aculeatus*, *Homalocladium platycladium*, *Asparagus cochinchinensis.* They are flat abnormal stem, and always degenerate to scale.

2.2.2 Shoot thorn Observing the thorn of *Gleditsia sinensis*, and to find their location. They form of axillary bud, and not easy to peel off.

2. 2. 3 Stem tendril　Observing the stem tendril of *Luffa cylindrica*, and to find their location.

3. Primary structure of stem

3. 1 Primary structure of dicotyledon　Bring cross section of stem of *Helianthus annuus* to distinguish epidermis, cortex and vascular cylinder under the lower microscope, and vascular bundle is round and pith rays lie on the bundles and pith is wide and large on the center, then turn to the higher microscope to observe from outer to inter (Figure 5 - 1).

3. 1. 1 Epidermis　It is a layer of slice cell in the outer, flat cells are arranged closely and regularly, and their outer wall are thick and cuticle, and can see nonglandular hair in sometime.

3. 1. 2 Cortex　It is layers of slice cells and has no intercellular spaces, and occupies a little part to compare with primary structure. Layers of cells are small near the epidermis and collenchymal. The cells are thick at the corner. The similar round chloroplasts are dyed green in the cells. There are stratas of slice cells in it, and they have a small secretory cavity. There aren't casparian strip inter the cell of the cortex.

3. 1. 3 Vascular cylinder　It occupies the wide space, and including the vascular bundle, pith ray and pith.

Vascular bundle: It consists of numbers of big or small open collateral bundles, and arranged a round. Every vascular bundle composes of primary phloem, cambium in the bundles and primary xylem.

Primary phloem lies out the vascular bundle, and its outer is poly - angle primary phloem fiber. The fiber is obviously thick, but not lignification, so dyed green. There are sieve tube, companion cell and slice cells of phloem in the primary phloem fibers.

Primary xylem includes protoxylem and metaxylem. The cross section ofvessels are similar round or poly - angle, and to distinguish protoxylem near the center according to big or small of mouth and thin or thick of dyed vessels. The mouth of vessels are small, come into early and dyed thicker; but metaxylem lies near the fascicular cambium, and the mouth of vessels are big, come into late and dyed thinner.

Pith rays are slice cells between two vascular bundle and relate in the cortex and out the pith. They have conduction and storage ability.

Pith lies on the center of stem, and is slice cells in the center of vascular cylinder. The cells are arranged loosely, and have storage ability.

4. The secondary structure of dicotyledon stem

Bringing cross section of 3 - 4 years' stem of *Tilia amurensis* from outer to inter to observe (Figure 5 - 2)。

4. 1 Periderm　It is obvious, including phellem、phellogen and phellodem.

4. 1. 1 Phellogen　It comes into cortex of storing divided ability, and cells are flat and cytoplasm is thick.

4. 1. 2 Phellem　It consists of the falt cells of arranged regularly on the radius, and cells wall are thick suberifacation, and cells are dead.

4. 1. 3 Phelloderm　It is 1 – 2 layers of cells, cytopasm is thick and it has nucleus and dyed deeply.

4. 2 Cortex　It is narrow, and composes of layers of parenchyma cells, and calcium oxalate cluster crystal can be observed in some cells.

4. 3 Phloem　Cells are arranged into ladder – shaped (near the cambium at the bottom) . It contributes alternation with parenchyma cells of pith ray arranged megaphone – shaped. It is obvious to contribute into horizontal dyed red phloem fibers alternation with dyed green parenchyma cells, sieve tube and companion cell of phloem. Primary phloem is already broken.

4. 4 Cambium　It consists of flat rectangle cells of arranged regularly and densely. It is cambium zone, and arranged into round.

4. 5 Xylem　It occupies a large part of space in the cambium on the cross section. Main is secondary xylem. A number of concentric circles can be observed, according to the big or small of diameter and thin or thick of wall of their cells, namely annual ring. What is different on the early wood and late wood of tissue structure? There are a lot of small vessels near the pith, named primary xylem.

4. 6 Pith　It lies on the centre of stem, and is composed of parenchyma cells, some has calcium oxalate cluster crystal, and some has slime or tannin. So a part of cells are dyed thicker.

4. 7 Pith ray　They send out radial from parenchyma cells of pith passed xylem and phloem to cortex. They are 1 – 2 lines cells in the xylem and amply to megaphone – shaped in the phloem.

4. 8 Vascular ray　Every vascular ray composes of parenchyma cells of conducting ability on the crosswise in the xylem and phloem. They are shorter than ray. They lie in the xylem named xylem ray and lie in the phloem named phloem ray.

5. The secondary structure of herbaceous stem of dicotyledon

Bringing cross section of stem of *Mentha haplocalyx* from outer to inter to observe.

5. 1 Periderm　It is composed of layers of arranged loose falt cells. and outer wall has thicker cuticle, and some have hair (grandular hair, nongranlulular hair or grandular scale).

5. 2 Cortex　It is norrow, and composed of flat cells. There are collenchyma composed of 10 layers thick wall cells, and the corners of cells are thick obviously. The cross section is dyed green. Endodermis is obvious and has red casparian dots on the radial walls.

5. 3 Vascular cylinder

5. 3. 1 Vascular bundle　It is composed of 4 bigger vascular bundles which opposite to the stem arris and 4 smaller vascular bundles between each arris, and they arrange into round. Phloem on the outer is narrower. Cambium is round, interfascicular cambium is obvious. Xylem is developed on the edges and corners. Vessels are single line, and numbers of rows, ar-

ranged on the radial direction. Vascular ray composed of parenchyma cells lie among the vessel.

5. 3. 2 Pith　It is developed and makes of big silce cells.

5. 3. 3 Pith rays　They are composed of parenchyma cells among vascular bundles and wide or narrow.

Moreover, some have fan – shaped radial grained hesperidin crystal in the every parenchyma cells of stem.

Observing cross section of stem of *Leonurus japonicus*. Its structure is similar with *Mentha haplpcalyx*, but there is calcium oxalate acicular crystal in the cortex cells, and pericycle fiber scattered inter the endodermis, calcium oxalate acicular and solitary crysta in the pith cells.

6. The secondary structure of gymnospermae stem

6. 1 Bringing cross section of 3 – 4 years'stem of *Pinus tabulaeformis* from outer to inter to observe.

6. 1. 1 Xylem　It hasn't vessels and typical xylem fibers, and mainly made of tracheid、parenchyma cells of xylem and xylem ray.

6. 1. 2 Phloem　It hasn't divided sieve tube and companion cell, and consists of sieve cell、parenchyma cells of phloem and phloem ray.

6. 1. 3 Resin cannal　There are lots of resin cannals in the cortex and vascular bundle.

6. 2 Radial section　Pay attention to arranged form of annual ring, sap wood, heart wood and ray.

6. 3 Tangential section　What is the difference of the upper part to compared with 6. 1, 6. 2?

Experiment report

1. Draw the outer formation of stem, and note the name of every section.

2. List the type of stem modifications.

3. Draw the dissect structure of cross section of *Helianthus annuus* stem, and note the name of every section.

4. Draw the dissect structure of stem of Tilia amurensis and *Pinus tabulaeformis*, and note the name of every section.

5. Draw the dissect structure of stem of *Mentha haplpcalyx*, and note the name of every section.

Questions

1. How to distinguish root and underground stem modifications?

2. What are the differences in the primary structure of dicotyledon root and stem?

3. How is secondary structure of dicotyledon stem formed?

4. What are the differences in the secondary structure of dicotyledon stem and gymnospermae stem?

实验六　单子叶植物茎和根状茎的构造以及双子叶植物根状茎的异常构造

【目的要求】

1. 掌握单子叶植物茎的构造，了解单子叶植物根状茎的构造。
2. 了解双子叶植物根状茎的异常构造（三生构造）。

【实验材料】

玉米茎的横切片；石斛茎横切片；石菖蒲根茎横切片；黄连根茎和大黄根茎的横切片；海风藤茎横切片；茅苍术根状茎的横切片。

【仪器用品】

显微镜。

【内容与方法】

一、单子叶植物茎的构造

观察玉米茎的横切片，由外向里观察。

1. 表皮　为茎的外层细胞，细胞排列整齐，扁长方形，外壁有较厚的角质层。

2. 基本组织　靠近表皮的数层细胞较小，排列紧密，胞壁增厚而木质化，形成厚壁组织，其内为薄壁组织，是基本组织系统的主要部分。

3. 维管束　分散在基本组织中，外方的维管束小，内方的渐大，没有皮层和髓的界限，也没有维管柱的界限。换高倍镜仔细观察一个维管束的结构，可见每个维管束外围有一圈由纤维组成的维管束鞘，里面只有初生木质部和初生韧皮部，其间没有形成层，是有限维管束。初生木质部中的导管在横切面上排成V字形，上半部是后生木质部，含有一对并列的大导管，下半部为原生木质部，有1～2个纵向排列的小导管、少量薄壁细胞和一个大空腔，大空腔是由于茎的伸长而将环纹或螺纹导管扯破形成。

初生木质部外方是初生韧皮部，其中原生韧皮部已被挤压破坏，后生韧皮部明显，通常只含有筛管和伴胞。

观察石斛茎的横切片，与玉米茎相似。

二、单子叶植物根状茎的构造

观察石菖蒲根状茎横切制片。

1. 表皮　为一层类方形表皮细胞，外壁增后，角质化。

2. 皮层　较宽广，散有油细胞、纤维束、叶迹维管束。纤维束类圆形，周围细胞中含有草酸钙方晶，形成晶（鞘）纤维；叶迹维管束外韧型，周围有维管束鞘包围。内皮层显著，具凯氏带加厚。有的有根迹维管束通过。

3. 维管束　内皮层以内的基本组织中，散有多数维管束，主要为周木型，靠内皮层排列较密，有的为外韧型。维管束鞘纤维发达，周围细胞中含有草酸钙方晶。

三、双子叶植物根状茎的构造

取黄连根状茎横切片，由外向内仔细观察（附图6－1）。

1. 木栓层　为数列木栓细胞。有的外侧附有鳞叶组织。

2. 皮层　宽广，石细胞单个或成群散在。有时可见根迹维管束斜向通过。

3. 维管束　为无限外韧型，环列，束间形成层不明显。韧皮部外侧有初生韧皮纤维束，其间夹有石细胞，切片染成鲜红色。木质部细胞均木化，包括导管、木纤维和木薄壁细胞。

4. 髓　由类圆形薄壁细胞组成。

取茅苍术根状茎横切片观察，与黄连相似，但薄壁细胞含草酸钙针晶和菊糖。

四、双子叶植物根状茎的异常构造

（1）观察大黄根状茎横切制片。在低倍镜下可见木质部和宽广的髓部。髓部有多数星点状的异型维管束。换高倍镜观察异型维管束，其形成层呈环状，内方为韧皮部，外方为木质部，射线成星状射出。

（2）观察海风藤的横切片，注意除正常的维管束18～33个排列成环外，在茎的中央髓部还有异常维管束6～13个，也排列成环。

【实验报告】

1. 列表比较双子叶植物根茎与单子叶植物根茎在解剖构造上的不同点。

	黄连根茎	石菖蒲根茎
保护组织		
皮层		
形成层		
维管束		
根迹、叶迹		
髓		

2. 绘黄连根茎横切面简图，注明各部位名称。
3. 绘玉米茎横切面简图，注明各部位名称。

【思考题】

大黄的异常构造与何首乌、牛膝的异常构造有何区别？

Experiment 6 Structure of Stem and Rhizome of Moncotyledon, Anomalous Structure of Rhizome of Dicotyledon

Aim and demand

1. Grasp the structure of monocotyledon stem and understand structure of its rhizome.

2. Understand anomalous structure of rhizome of dicotyledon (tertiary structure).

Experiment materials

Cross section of stem of Yumi (*Zea mays* L.), Shihu (Dendrobium nobile Lindl.); cross section of rhizome of Shichangpu (*Acorus tatarinowii* Schott.); cross section of rhizome of Huanglian (*Coptis chinensis* Franch.) and Dahuang (*Rheum officinale* Baill.); stem of Haifengteng (*Piper kadsura* (Choisy) Ohwi.); Changzhu (*Atractylodes lancea* (Thunb.) DC.).

Instruments and appliances

Microscope.

Contents andprocedures

1. The structure of stem of monocotyledon

Bringing cross section of stem of *Zea mays* from outer to inter to observe .

1.1 Periderm It is composed of rectangle regularly parenchyma cells at the outer, and the outer wall have thicker cuticle.

1.2 Ground tissue Layers of cells near the periderm are arranged closely, and cell walls are thick and suberifacation, and form the sclerenchyma. There are parenchyma tissue in it, and they are the main part of ground tissue.

1.3 Vascular bundle They are dispersed into the ground tissue, vascular bundle are smaller near the outer, and bigger near the inter. They haven't border of cortex, pith and vascular cylinder. Then turnning to the higher microscope to observe a vascular bundle. There is a round vascular sheath composed of fiber out of the vascular bundle. They have only primary xylem and phloem, and haven't cambium among them, and are a closed collateral bundle. Vessels of primary xylem arrange "V" on the cross section. Protoxylem lies on the upper, and have a big stand side by side vessel. Metaxylem lies on the under, and 1 – 2 radial arranged small vessels, a little parenchyma cells and a hole. The hole forms of divided annular and spiral vessel because of extended stem.

Primary xylem lies in the primary phloem, and protophloem is already damaged. Metaphloem is obvious and has only sieve tube and companion cell.

Observing cross section of stem of *Dendrobium nobile*, you may find similarities with Zea mays.

2. The structure of rhizome of monocotyledon

Observingcross section of stem of *Acorus tatarinowii.*

2.1 periderm It is composed of similar rectangle cells, and the outer wall are thicker and cuticle.

2.2. Cortex It is wide, and dispersed in the oil cell, fiber bundle and leaf scar and root scar vascular bundle. Vascular bundle are similar round, and have calcium oxalate crystal among the cells, and they form crystal fibre. Leaf scar vascular bundles are collateral and vascular sheath among them. The endodermis is obvious and has a thick casparian strip. Some have root scar vascular bundle.

2.3 Vascular bundle Dispersed numbers of vascular bundle lie in the ground tissue of the endodermis, and are mainly amphivasal. They are only arranged closely near the endodermis, and some are collateral. The fibers of vascular sheath are developed, and have calcium oxalate crystal among the cells.

3. The structure of rhizome of dicotyledon plant

Bringing cross section of rhizome of *Coptis chinensis* from outer to inter to observe (Figure 6-1).

3.1 Phellem It consists of the suberifacation cells, and some have tissue of leaf scale.

3.2. Cortex It is wide, and dispersed into single or cluster stone cells, and some have inclined root scar.

3.3. Vascular bundle It is open collateral bundle, and arrange into round, and interfascicular cambium isn't obvious. There are fiber bundle of primary phloem in the outer, and disperse into stone cell among it. Cross section is dyed red. Xylem cell are lignification, and include vessel, fiber of xylem and parenchyma cell of xylem.

3.4 Pith It is composed of similar round parenchyma cells.

Bringing slide of cross section of rhizome of *Atractylodes lancea* to observe and find that it is similar with Chinese goldthread, but it has calcium oxalate acicular crystal and inulin in the parenchyma cells.

4. The anomalous structure of rhizome of dicotyledon

4.1 Observing cross section of rhizome of *Rheum officinale* under the lower microscope, to find xylem and wide pith. The pith has numbers of star-shaped anomalous vascular bundle. Then turn to the higher microscope to observe anomalous vascular bundle. Its cambium is round, and phloem at the inter and xylem at the outer and ray send out star-shaped.

4.2 Observing cross section of *Piper kadsura*, to pay attention to except normal 18 ~ 33 arranged ring vascular bundles, and 6 - 13 abnormal vascular bundles at the centre pith of stem, and arranged ring, too.

Experiment report

1. Compare the structure of rhizome of dicotyledon with monocotyledon different on the dissect structure ?

	rhizome of *Coptis chinensis*	rhizome of *Acorus tatarinowii*
protective tissue		
cortex		
cambium		
vascular bundle		
leaf scar、root scar		
pith		

2. Draw the dissected structure of *Coptis chinensis*, and note the name of every section.
3. Draw the dissected structure of *Zea mays*, and note the name of every section.

Question

What are the differences between the anomalous structure of rhizome of *Rheum officinale*, *Polygonum multiflorum* and *Achyranthus bidentata*.

实验七　叶的形态和内部构造

【目的要求】

1. 识别掌握叶的外部形态、类型及变态叶的形态、特征和类型。
2. 掌握叶的组成部分，叶片的形状及叶端、叶基、叶缘；叶的分裂、叶脉的类型；单叶与复叶的区别；复叶的类型；叶序。
3. 掌握双子叶植物叶的解剖构造；了解单子叶植物叶的构造。

【实验材料】

温室中的鲜活植物：扶桑、木槿、紫草、忍冬、三棵针、黄精、银杏、薄荷的茎枝；蒲公英、马蹄莲、龟背竹的花序；洋葱、百合、麻黄的茎；仙人掌、三棵针、枸骨、洋槐的叶刺与托叶刺；豌豆及菝葜的卷须；蜡叶标本：月季、槐树或紫藤、合欢、南天竹、酢浆草、白花草木樨、刺五加、甘草、葛藤、酸橙等植物的叶；薄荷叶横切片；淡竹叶横切片。

【仪器用品】

镊子、放大镜、解剖镜、解剖用具、显微镜。

【内容与方法】

一、叶的形态

1. 叶的组成　观察扶桑、木槿等植物叶的形态，分辨叶片、叶柄、托叶；注意其叶端、叶基、叶缘的形态和脉序的类型。

2. 叶片的形态　观察蜡叶标本。

3. 单叶和复叶

（1）单叶和复叶的区别 观察月季、槐树或紫藤的羽状复叶，与扶桑、木槿叶比较。注意观察复叶的小叶柄与总叶柄之间没有腋芽，而总茎轴之间和单叶的叶柄与茎柄之间一样，均具有腋芽。

（2）复叶的类型 观察合欢、紫藤、南天竹、酢浆草、白花草木樨、五加、甘草、望江南、葛藤、酸橙等植物的叶，判断其为何种复叶。

4. 叶序 观察紫草、忍冬、三棵针、黄精、银杏、薄荷的茎枝，判断其叶序类型。

5. 叶的变态

（1）苞叶 观察蒲公英、马蹄莲、龟背竹的花序，指出哪一部分是叶的变态。

（2）鳞叶 观察洋葱、百合、麻黄的茎，找出变态的叶，区别肉质鳞叶与膜质鳞叶。

（3）叶刺 观察仙人掌、三棵针、枸骨、洋槐等植物，注意区别叶刺与托叶刺。

（4）卷须叶 观察豌豆由复叶顶端二、三对小叶变成的卷须及菝葜由托叶变成的卷须。

二、叶的解剖构造

1. 双子叶植物叶片的构造 观察薄荷叶的横切制片（附图 7－1）。

（1）表皮 上表皮细胞长方形，下表皮细胞较小，扁平，均被角质层，有气孔；表皮外有腺鳞、小腺毛和非腺毛。

（2）叶肉 栅栏组织为一层细胞，海绵组织为 4～5 层排列疏松的薄壁细胞。

（3）主脉 维管束外韧型，木质部位于主脉的近轴面（靠近上表皮），导管常 2～5 个纵列成数行，韧皮部位于木质部下方，较窄，细胞小，形成层明显。主脉上、下表层内侧常有厚角组织。

表皮细胞、薄壁细胞中常含有橙皮苷结晶，呈针簇状。

2. 单子叶植物（禾本科）叶片的构造

观察淡竹叶的叶片横切制片（附图 7－2）。

（1）表皮 上表皮细胞类方形，大小不一，壁薄。大型表皮细胞（泡状细胞或运动细胞）成扇形，下表皮细胞较小，排列整齐。上、下表皮均被角质层、气孔及单细胞非腺毛。

（2）叶肉 栅栏组织和海绵组织分化不明显。上表皮下方有一列短圆柱形薄壁细胞，内含叶绿体，并通过主体，呈栅栏组织状。

（3）主脉 上部向下微凹，下部向外突起。维管束外韧型，无形成层，周围有 1～2 列纤维组成的维管束鞘包绕，木质部导管稀少，排成 V 形，其下方为韧皮部。在上、下表皮的内侧有厚壁纤维群。

【实验报告】

1. 绘一片完全叶的外形图，注明各部分名称。
2. 绘薄荷叶横切面简图，注明各部分名称。
3. 把观察到的变态叶及叶序类型列表记录下来。

【思考题】

1. 如何区分单叶与复叶？常见的复叶有哪几种类型？
2. 比较单子叶植物与双子叶植物的叶在构造上的异同点？

Experiment 7 Morphology and Internal Structure of Leaf

Aim and demand

1. Gain the outer morphology and type of leaf and morphology, type, character of leaf modifications.

2. Gain the composition and shape of leaf, and leaf apex, leaf base, leaf margin, division of leaf, type of vein, the distinction between simple leaf and compound leaf, type of compound leaf, and phyllotaxy.

3. Gain the structure of dicotyledon leaf ; understand structure of monocotyledon leaf.

Experiment materials

Living plant in greenhouse: leaf of Fusang (*Hibiscus rose – sinensis* L.) and Mujin (*Hibiscus syriacus* L.), the stem of Zicao (*Ligustrum obtusifolium* Sieb. et Zucc.), Rendong (*Lonicera japonica* Thunb.), Sankezhen (*Berberis julianae* Schneid.), Huangjing (*Polygonatum sibiricum* Delar. ex Red), Yinxing (*Ginkgo biloba* L.), Bohe (*Mentha haplocalyx* Briq.); the inflorescence of Pugongying (*Taraxacum mongolicum* Hand. – mazz.), Matilian (*Zantedeschia aethiopica* (L.) Spreng.), Guibeizhu (*Monstera diliciosa* Liebm.); the inflorescence of yangcong (*Allium cepa* L.), baihe (*Lilium Brounii* F. E. Brown. var. *viridulum* Backer.), Mahuang (*Ephedra sinica* Stapf); Xianrenzhang (*Opuntia dillenii* (Ker – Gael.) Haw.), Gougu (*Ilex cornuta* Lindl.), Yanghuai (*Sophora japonica* L.); leaf tendril of Wandou (*Pisum sativum* L.) and rhizome of Baqia (*Smilax sieboldii* Miq.) . Wax sample: leaves of Yueji (*Rose chinensis* Jacq.), Huaishu (*Sophora japonica* L.), Ziteng (*Wistaria sinensis* (Sims.) Sweet.), Hehuan (*Albizzia julibrissin* Durazz.), nantianzhu (*Nandina domestica Thunb*.), Zhajiangcao (*Oxalis corniculata* L.), Baihuacaomuxi (Melilotus albus Desr.), Ciwujia (*Acanthopanax senticosus* (Rupr. et Maxim.) Harms. , Gancao (*Glycyrrhiza uralensis* Fisch.), Geteng (*Pueraria lobata* (Willd.) Ohwi), Suancheng (*Citrus aurantium* L.); cross section of Bohe (*Mentha haplocalyx* Briq.); cross section of Danzhuye (*Phyllostachys nigra* (Lodd.) Munro var. *henonis* (Mitf.) stapf ex Rendle.).

Instruments and appliances

Tweezers, magnifying glass, anatomical lens, dissector microscope, microscope.

Contents and procedures

1. The outer morphology of leaf

1.1 The composition of leaf Observing the outer morphology of leaves of *Hibiscus rose – sinensis*. *Hibiscus syriacus* and others, distinguishing blade, petiole and stipule; Pay attention to their formation of leaf apex、leaf base、leaf margin and type of vein.

1. 2 The morphology of blade Observing wax sample.

1. 3 Simple leaf and compound leaf

1. 3. 1 The division of between simple leaf and compound leaf Observing pinnately compound leaf of *Rose chinensis*, *Sophora japonica*, or *Wistaria sinensis* . Comparing with *Hibiscus rose - sinensis* and *Hibiscus syriacus*, noting that there is no axillary bud between little petiode and total petiode of compound leaf , but they all have axillary bud the same among stem as petiole and stem.

1. 3. 2 Type of compound leaf Judging their type of compound leaf , through observing the leaves of *Albizzia julibrissin* , *Wistaria sinensis*, *Nandina domestica*, *Oxalis corniculata*, *Melitotus albus*, *Acanthopanax senticosus*, *Glycyrrhiza uralensis*, *Pueraria lobata*, and *Citrus aurantium* .

1. 4 Phyllotaxy Judging their type of phyllotaxy through observing the stems of *Lithospermum*, *Lonicera japonica chryasantha*, *Berberis julianae*, *Polygonatum sibiricum*, *Ginkgo biloba*, and *Mentha haplocalyx* .

1. 5 Leaf modifications

1. 5. 1 Bract Observing the inflorescence of *Taraxacum mongolicum*, *Zantedeschia aethiopica*, and *Monstera diliciosa* and pointing out where the leaf modification is.

1. 5. 2 Scale leaf Observing the inflorescence of *Allium cepa* , *Lilium pumilum*. Var. *viridulum*, *Ephedra sinica* and pointing out abnormal leaf and comparing with membranaceous and succulent.

1. 5. 3 Thorn leaf Observing *Opuntia dillenii*, *Berberis julianae*, *Ilex cornuta and Sophora japonica* distinguishing thorn leaf and thorn stipule.

1. 5. 4 Leaf tendril Observing 2 - 3 little leaf on the tip of compound leaf of *Pisum sativum* , they change into leaf tendril, and the stipule turn into leaf tendril of rhizoma Smilax sieboldii.

2. The dissected structure of leaf

2. 1 The structure of dicotyledone done Observing cross section of leaf of *Mentha haplocalyx* (Figure 7 - 1).

2. 1. 1 Epidermis Upper strata cell is rectangle, the lower layer is small and flat, and is covered with cuticle; and stonata, outer epidermis have glandular hair and no glandular hair.

2. 1. 2 Mesophyll Palisade tissue is a layer of cell and spongy tissue is 4 - 5 layers loose and parenchyma cells.

2. 1. 3 Vein The vascular bundle is collateral , the xylem lie near axle of vein (near upper epidermis), the vessel often split Vertically into 2 - 5 vessels, the phloem under the xylem. It is narrower and smaller cell. Its cambium is obvious. There are often collenchymas tissue in the upper and lower layer of vein.

There are the acicular or cluster crystal ofhesperidin, inepidermis and slice cell.

2. 2 The dissect structure of monocotyledon (Gramineae) Observing cross section of leaf of *Phyllostachys nigra* var. *henonis* (Figure 7－2).

2.2.1 Idermis Upper layer cell is similar rectangle, big or small and parenchymas cells. The big epidermis cell (bulliform cell or motor cell) is fan－shaped. The lower epidermis cell is smaller and regularly arranged. Upper strata and lower all covered by cutide stoma and no glandular hair.

2.2.2 Mesophyll Palisade and spongy tissue aren't obvious. There are cutide, stoma and noglandular hair on upper epidermis and lower.

2.2.3 Vein The higher is concave and the bellower is raised. The vascular bundle is collateral and doesn't have the cambium. Around the vascular bundles, there are vascular bundle which are made up of 1～2 range of fibers. The line of the xylem are few and a arrange of "V", the phloem under it. There are sclerenchyma in the upper and lower layer of vein.

Experiment report

1. Draw the outer formation of a complete leaf, and note the name of every section.
2. Draw the dissected structure of *Mentha haplocalyx*, note the name of every section.
3. List leaf modifications and the type of phllotaxy.

Questions

1. How to distinguish between simple leaf and compound leaf? How many types are compound leaf divided?
2. Compare differences the dissected structure of dicotyledone and monocotyledon.

实验八　花的形态和内部构造

【目的要求】

1. 识别掌握被子植物花的组成部分和一般内部构造；学会解剖花以及使用花程式描绘花的方法。
2. 理解花序的概念并掌握常见花序类型的特征。
3. 识别掌握被子植物花的子房位置及胎座类型。
4. 了解花粉粒的显微形态特征。

【实验材料】

花及花序的浸泡标本：龙牙楤木、山楂、车前、马鞭草、杨、龟背竹、马蹄莲、绣线菊、小茴香、白芷、蒲公英、无花果、附地菜、紫草、香雪兰、射干、卫茅、细叶百合、杏、桔梗、南瓜、石竹、大戟、益母草。

温室中的鲜活植物：白菜、荠菜、葶苈、豌豆、黄瓜。

柿蒂的横切制片，月季的横切制片；松花粉、蒲黄、百合、芍药、南瓜等的花粉制片。

【仪器用品】

放大镜、解剖镜、解剖用具、镊子、显微镜。

【内容与方法】

一、花的组成

取一朵白菜花，用解剖针和镊子轻轻地分别由下向上或由外向里逐层剥离，按顺序将各部分平放在白纸或培养皿上；注意观察各部分的形态特征和排列情况并写出它的花程式。

另取 2 ~3 种不同类型的花，用上述方法进行观察，分析其不同的组成，并写出各自的花程式。

二、花序的类型

1. 无限花序 观察鲜活植物白菜、荠菜、葶苈的花序，可见其花序轴较长，没有分枝；小花具花柄，花柄近等长；小花的开放顺序是由下向上依次进行的，而花序轴顶端仍保持生长的能力，这种花序可称为总状花序。

观察龙牙楤木、山楂、车前、马鞭草、杨、龟背竹、马蹄莲、绣线菊、小茴香、白芷、蒲公英、无花果等植物的花序，注意与总状花序有何异同点，指出其各自属于哪种花序。

2. 有限花序 观察附地菜、紫草的花序，注意其顶端首先开放，因此花序轴不能继续生长，只是在其下方产生一侧轴，接续生长而成。由于其花序轴下分枝均向同一侧方向，故称螺旋状聚伞花序。

继续观察香雪兰、射干、卫茅、石竹、大戟、益母草等植物的花序，注意与螺旋状聚伞花序的异同点，写出其花序类型。

三、花的内部构造

1. 萼片的内部构造 观察柿蒂的横切制片，注意萼片和叶片的内部构造有何相似点，柿的萼片表皮有无毛茸和毛茸的类型。上下表皮之间有无栅栏组织和海绵组织的分化，有无厚壁组织和草酸钙结晶。

2. 花瓣的内部构造 观察月季的花瓣制片，注意上表皮细胞有何特征，有无栅栏组织和海绵组织的分化。

四、子房的位置

1. 子房上位、花下位或花周位 观察白菜、细叶百合、杏的花，注意子房与花托的愈合程度，并注意区分花下位与花周位两种类型。

2. 子房下位（中位）、花周位 观察桔梗、南瓜的花，注意雌蕊的哪些部分独立露出。

3. 子房下位、花上位 观察黄瓜的花，注意是否还能观察到子房。

五、胎座的类型

取百合的子房作横切，在放大镜或解剖镜下进行观察：可见子房壁内有 3 个子房室，每两个子房室之间的部分是2个心皮的结合处，是一隔膜，外侧有一凹陷，即腹缝线，其内有一堆维管束，是两个心皮侧束合并而成；在每个心皮中央有一中脉维管束（背束），与其相应的凹陷即背缝线，3个心皮内卷形成中轴，胚珠着生于腹缝线愈合形成的隔膜上，故称中轴胎座。

另取豌豆、桔梗、石竹、黄瓜等植物的子房做横切，注意观察其子房的室数、心皮的数目及胚珠的着生情况，指出其各属于哪种类型的胎座。

六、花粉粒的显微特征

取1~2种花的花粉少许，用水合氯醛液透化后，在显微镜下观察。注意花粉形态、大小、类型，如单粒或复合；有无萌发孔、萌发孔的形状、位置、数目；表面特征，如光滑或呈波浪形，有由小刺、瘤、颗粒等形成的各样雕纹。

【实验报告】

1. 任绘一种植物的花，注明各部名称；写出该植物及其他两种植物花的花程式。
2. 绘总状花序、穗状花序、伞形花序、伞房花序、头状花序、二歧聚伞花序的简图。
3. 绘柿蒂的萼片的构造图，标明各部分的名称。
4. 绘各种胎座横切面的简图。

【思考题】

1. 为什么说花是一个适应生殖功能的变态枝条？
2. 何谓花序？如何区别有限花序和无限花序？
3. 雌蕊是由什么构成的？分几部分？怎样判断组成雌蕊的心皮数？
4. 花萼、花瓣和叶片的显微构造有何相似点？

Experiment 8　Morphology and Internal Structure of Flower

Aim and demand

1. Grasp the composition and the dissect structure of flower , learn to dissect flower and use flower formula to describe a flower.

2. Understand the concept of inflorescence and gain the characters of common type of inflorescence.

3. Grasp the living location of the ovary and the type of placenta of angiospermae.

4. Understand the formation character of pollen grain.

Experiment materials

Immersioning sample of flower and inflorescence: the inflorescence of Longyacongmu (Aralia elata (Miq.) Seem.), Shanzha (*Cratagus pinnatifida* Bunge.), Cheqian (*Plantago asiatica* L.), Mabiancao (*Verbena offcinalis* L.), Yang (*Populus cathayana* Rehd.), Guibeizhu (*Monstera diliciosa* Liebm.), Matilian (*Zantedeschia aethiopica* (L.) Spreng.), Xiuxianju (*Spiraea salicifolia* L.), Xiaohuixiang (*Foeniculum vulgare* Mill.), Baizhi (*Anegelica dahurica* (Fish. ex Hoffm.) Benth et Hook. f.), Pugongying (*Taraxacum mongolicum* Hand. – Mazz.), Wuhuaguo (*Ficus craica* L.), Fudicai (*Trigonotis peduncularis* (Tev.) Benth. Ex Baker et Moore), Zicao (*Lithospermum erythrorhizon* Sieb. Et Zucc.), Xiangxuelan (*Freesia refracta* Klatt.), Shegan (*Belamcanda chinensis* (L.) DC.), Weimao (*Euonymus alatus* (Thunb.) Sieb.), Xiyebaihe (*Lilium pumilum* DC.), Xing (*Prunus armeniaca* L.), Jiegeng (*Platycodon grandiflorum* (Jacq.) A. DC.), Nangua (*Cucubita moschata* (Duch.) Doiret.), Shizhu (*Dianthus chinensis* L.), Daji (*Euphorbia pekinensis* Rupr.), Yimucao (*Leonurus heterophyllus* Sweet.).

Living plant of greehouse: inforescence of Baicai (*Brassica pekinensis* (Lour.) Rupr.), Jicai (*Capsella bursa – pastoris* (L.) Medic.), Tingli (*Draba nemorosa* L.), Wandou (*Pisum sativum* L.), Huanggua (*Cucumis sativious* L.) .

Slide of crosssection of sepal Shi (*Diospyros kaki* Thunb.) And Yueji (*Rosa chinensis* Jacq.); pollen of Yousong (Pinus tabulaeformis Carr.), Puhuang (*Typha angustifolia* L.), slide of pollen Baihe (*Lilium brownie* F. E. Brown. Var. viridulum Backer), Shaoyao (*Paeonia lactiflora* Pall.) and Nangua (*Cucurbita moschata* (Duch.) Doiret) .

Instruments and appliances

Tweezers, magnifying glass, anatomical lens, dissector , microscope.

Contents and procedures

1. The composition of flower

Bring a flower of *Brassica pekinensis*, and use an anatomic needle and tweezers to peel off flower one by one from the upper to the lower or from the outer to the inter, put every section on the newsprint or the culture dish, according to order observing the formation and the arrangement of every section, and write down the flower formula.

Moveover, binging 2 – 3 other types of flower, and applying above – mentioned to observe and analyse their different composition, and describing every flower formula.

2. The type of inflorescence

2.1 Indefinite inflorescence Observing the inflorescence of *Brassica pekinensis*, *Capsella bursa – pastoris*, *Draba nemoros*. It is thus clear that their rachis are longer, aren't branch; the little flower has a flower stalk, and the flower stalks are almost equal in length; coming into flower of the little flowers are from the lower to the upper, and the tip of the rachis keeps the capability of the growing, this inflorescence is names the raceme.

Observing the inflorescence of *Aralia elata* , *Cratagus pinnatifida*, *Plantago asiatica* , *Verbena offcinalis*, *Populus cathayana*, *Monstera diliciosa*, *Zantedeschia aethiopica*, *Spiraca salicifolia*, *Foeniculum vulgare*, *Anegelica dahurica*, *Taraxacum mongolicum*, *Ficus craica*, ect. Paying attention to the difference with the raceme, then to decide the type of their inflorescence.

2.2 Difinite inflorescence Observing the inflorescene of *Trigonotis peduncularis*, *Lithospermum erythrorhizon*. It is clear that the upper flowers bloom first, thus the rachis can not continue growing any more, and have to produce a branch to continue growing. The branch of the rachis grow towards the same direction, so they are named bostrix.

Observing the inflorescence of *Freesia refracta*, *Belamcanda chinensis*, *Euonymus alatus*, *Dianthus chinensis*, *Euphorbia pekinensis*, *Leonurus heterophllus*, and comparing with bostrix, then to write down the type of inflorescence.

3. Internal structure of flower

3.1 The structure of calyx Observing the cross section of *Diospyros kaki*, and paying attention to similar dots of structure of calyx and leaf. Whether there has hair on the epidermis of calyx of *Diospyros kaki* or not? What type of hair? Are there the divided of palisade tissue and spongy tissue between upper and lower epidermis? Whether it has sclerenchyma and calcium oxalate crystal or not?

3.2 The structure of flower Observing the cross section of flower of *Rosa chinensis*, paying attention to what character is upper epidermis? Are there the division of palisade tissue and spongy tissue ?

4. The living location of ovary

4.1 Superior ovary, hypogynous flower or perigynous flower Observing the flowers of *Brassica pekinensis*, *Lilium pumilum*, *Prunus davidiana*, to pay attention to ally with ovary and receptacle, and distinguish the type of hypogynous flower and perigynous flower.

4.2 Inferior ovary (perigynous flower) Observing the flower of *Platycodon grandiflorum*, *Cucubita moschata*, to pay attention to which section the pistil exposed.

4.3 Inferior ovary, epigynous flower Observing the flower of *Cucumis sativious* to pay attention to whether the ovary. can be observed.

5. The type of placenta

Bringing the ovary of Baihe (*Liliumbrownie* F. E. Brown. Var. viridulum Backer) to cross cut and put down magnifying glass or anatomical lens to observe: It is clear that there are three ovary room, the part of two ovary room jioned of two carpels in ovary wall. It is a slider, there is a hollow out the outer, namely stomach thread. There is a pile of vascular hundle and they are alled with the branch of the carpels. There is a middle veration vascular hundle in the every middle carpel, and corresponding with hollow is back thread. The uxile is composed with 3 carpels. Ovule lie on slider composed of alled stomach thread, so it is named axile placenta.

Moveover, bringing ovary of Huanggua (*Cucumissativious* L.), Jiegeng (*Platycodongrandiflorum* (Jacq.) A. DC.), Shizhu (*Dianthus chinensis* L.), Wandou (*Pisumsativum* L.) tocrosscut and observe their room ofovarys, numbers of carpel and located situation of ovule, point out the type of their placenta.

6. The formation character of pollen grain

Bringing a little of pollen grain of 1 – 2 kinds of flower, using chloral hydrate to penetrate, put it on microscope to observe. Paying attention to the shape, size and type of pollen grain, single or compound, whether it has germind aperture or not, and the shape, position, number of germinal aperture; surface characters: such as smooth or wave, whether it has all kind of the carving in composition of the thorn, tumor or grain, and so on.

Experiment report

1. Draw the outer formation of a flower, and note the name of every section; writing down the flower formula of it and other two plants.

2. Draw a sketch of the inflorescence of raceme、spike、umbel、corymbs、capitulum and dichasium.

3. Draw the structure of *Diospyros kaki*, and note the name of every section.

4. Draw cross sections of all kinds of placentation.

Questions

1. Why the flower is an branch modification of a stem for fitting reproduction.

2. What is inflorescence? How to distinguish difinite and indefinite infloreescence?

3. What is composes of pistil? How many part can be divided? How to judge the number of carpels of pistil?

4. What are the similar points among structures of calyx, petal and leaf?

实验九 果实、种子的形态、类型和构造

【目的要求】

1. 识别掌握各种果实的类型和主要形态特征。
2. 掌握种子的外部特征和种子的组成部分，了解果实的内部构造。
3. 能够区别有胚乳种子和无胚乳种子。
4. 了解果实和种子的内部构造。

【实验材料】

各种干燥或浸制的果实标本：萝藦、草决明、大豆、荠菜、紫丁香、北马兜铃、虞美人、穿龙薯蓣、石竹或麦仙翁、曼陀罗、牛蒡、向日葵、玉米、板栗、榛、小茴香或白芷、杜仲、糖槭、芍药或牡丹、八角茴香、北五味子、白头翁、铁线莲。

鲜活植物果实和种子：番茄、枸杞、橘、黄瓜、山毛桃或榆叶梅、梨、草莓、桑、凤梨、无花果；蓖麻种子、杏仁、落花生；枸杞的横切制片、杏仁种子横切片。

【仪器用品】

放大镜、镊子、解剖镜、解剖用具、显微镜。

【内容与方法】

一、单果

1. 肉质果 取番茄、枸杞、橘、黄瓜、山毛桃或榆叶梅、梨的果实横切，注意其外、中、内各层果皮有无明显界限？质地如何？子房室数和胎座的类型？并分辨哪些属于真果，哪些是假果。

2. 干果

（1）取萝藦、草决明、大豆、荠菜、紫丁香、北马兜铃、穿龙薯蓣、石竹或麦仙翁、曼陀罗的果实进行观察，注意其成熟后的开裂方式，是一个缝线开裂或两个缝线开裂，是室间开裂或室背开裂或室轴开裂，并注意分辨其心皮数目、子房室数及胎座类型。

（2）取牛蒡、向日葵、玉米、板栗、榛、小茴香或欧白芷、杜仲、糖槭的果实观察，注意根据其果皮的质地、果皮与种皮分离的难易及果皮是否延伸成翅状等特征，区别不同类型的闭果。

二、聚合果

观察植物芍药或牡丹、八角茴香、北五味子、白头翁、草莓的果实，注意它们是由多个离生心皮形成的果实聚生于花托上，指出其各属于哪种类型的聚合果。

三、聚花果

观察桑、凤梨、无花果的果实，注意分辨其各自的食用部分为哪些结构？指出它们属于真果还是假果？

四、果实的内部构造

取宁夏枸杞横切片在显微镜下观察。

1. 外果皮　为一列扁平细胞，壁较薄，外被角质层，外缘作细齿状突起。

2. 中果皮　为10余列薄壁细胞，外侧1～2列较小，中部细胞较大，有的细胞含草酸钙砂晶；维管束双韧型，散列。

3. 内果皮　为一列椭圆形细胞，切向延长。

4. 种皮　最外为一列石细胞，类长方形，侧壁及内壁呈U字形增厚。其下为3～4列被挤压的薄壁细胞。最内一层为扁平长方形薄壁细胞，微木化。

5. 胚乳及胚根、子叶薄壁细胞　含有脂肪油和颗粒状物。

五、有胚乳种子和无胚乳种子

1. 有胚乳种子　观察蓖麻种子，呈扁平广卵形，一面较平，另一面较隆起。种皮坚硬，由3层结构构成，最外一层为膜质，具褐黑色花纹，有光泽；中层骨质，含黑褐色色素；内层为白色膜质。从种子表面观察，在种子较狭的一端有一浅色的海绵状突起，即种阜，能吸水，有利于种子萌发。位于种子腹面种阜内侧的小突起，即种脐。种阜和种脐的下方有一条纵向的向上的隆起为种脊，种脊的终点为合点。种孔被种阜遮盖，一般看不见。小心剥去种皮，其内肥厚的部分为胚乳，用刀片平行于种子的宽面作纵切，把胚乳分为两半，用放大镜观察，能见到两片大而薄且叶脉清晰的子叶，同时可看到胚根、极小的胚芽和很短的胚轴。

2. 无胚乳种子　取花生种子观察，种子长卵圆形。种皮薄，呈粉红或紫红色。剥去种皮，其内为两片肥厚、乳白色、有光泽的子叶。打开两片子叶，可观察到其胚芽是由一个主芽和两个侧芽组成。花生的胚轴位于胚芽下端，较粗壮，子叶着生于此。胚轴下方是胚根，突出于两片子叶之外，呈短喙状。

另外取蚕豆、杏仁进行观察，注意分辨其外部特征及内部构造。

六、种子的内部构造

取杏仁横切片观察。

1. 种皮外表皮 为一列薄壁细胞组成，散生长圆形、卵形的橙黄色石细胞，上半部突出于表面，下半部埋在薄壁组织中。

2. 种皮内表皮 为一列薄壁细胞，含黄色物质。

3. 外胚乳 为数列颓废的薄壁组织。

4. 内胚乳 为一列长方形细胞，内含糊粉粒及脂肪油滴。

5. 子叶细胞 为多列多角形薄壁细胞，含糊粉粒，较大的糊粉粒中有1细小草酸钙簇晶，并含脂肪油滴。

【实验报告】

1. 列表写出所观察到的果实类型，区别真果与假果。
2. 列表比较蓇葖果、荚果、角果、蒴果的主要区别。

	蓇葖果	荚果	角果	蒴果
心皮数目				
子房室数				
胎座类型				
开裂方式				

3. 绘宁夏枸杞果实的横切片简图，标明各部分的名称。
4. 绘蓖麻种子的外形图及剖面图，并注明各部分名称。
5. 绘杏仁外形图及横切面详图，标明各部分的名称。

【思考题】

1. 一般的果实是怎样发育形成的？
2. 果实的分类依据是什么？
3. 怎样区分有胚乳种子和无胚乳种子？
4. 通过实验怎样理解胚是一个幼小的植物体？

Experiment 9 Formation，Type and Structure of Fruit and Seed

Aim and demand

1. Gain the type and main formation characteristics of fruit.
2. Grasp the formation and structure of seed.
3. Distinguish between albuminous seed and exalbuminous seed.
4. Understand the structure of fruit and seed.

Experiment materials

Dried orImmersioning sample of flower : the fruit of Luomo (*Metaplexis* (Thumb.) Makino), Caojueming (*Cassia tora* L.), Dadou (*Glycine max* (L.) Merr.), Jicai (*Capsella bursa - pastoris* (L.) Medic.), Zidingxiang (*Syringa oblata* Lindl.), Beimadouling (*Aristolochia contorta* Bunge), Yumeiren (*Papaver rhoeas* L.), Chuanlongshuyu (*Dioscorea nipponica* Makino), Shizhu (*Dianthus chinensis* L.) or Maixianweng (*Agrostema githago* L.), Mantuoluo (*Datura stramonium* L.), Niubang (*Arctium lappa* L.), Xiangrikui (*Helianthus annuus* L.), Yumi (*Zea mays* L.), Banli (*Castanea mollissima* Blume), Zhen (*Corylus heterophylla* Fisch. ex Trautv.), Xiaohuixiang (*Foeniculum vulgare* Mill.) or Baizhi (*Angelica dahurica* (Fish. Ex Hoffm.) Benth et Hook. f.), Duzhong (*Eucommia ulmoides* Oliv.), Tangqi (*Acer negundo* L.), Shaoyao (*Paeonia lactiflora* Pall.) or Mudan (*P. suffruticosa* Andr.), Bajiaohuixiang (*Illicum verum* Hook. f.), Beiwuweizi (*Schisandra chinensis* (Turcz.) Baill), Baitouweng (*Pulsatilla chinensis* (Bunge) Regel), Tiexianlian (*Clematis hexapetala* Pall.).

Living plant of greehouse: Fanqie (*Lycopersicon esculentum* Mill.), Gouqi (*Lycium chinensis* Mill.), Ju (*Citrus reticulata* Bianco.), Huanggua (*Cucumis sativious* L.), Shanmaotao (*Prunus davidiana* (Carr.) Franch.) or Yuyemei (*Prunus triloba* Lindl.), Li (*Pyrus* Sp.), Caomei (*Fragaria ananassa* Duch.), Sang (*Morus alba* L.), Fengli (*Aechmea weilbachii* Didr.), Wuhuaguo (*Ficus carica* L.), The seed of Bima (*Ricinus communis* L), Xingren (*Prunus armeniaca* L.), Luohuasheng (*Arachis hypogaea* L.); Cross section of Gouqi (*Lycium chinense* Mill.), Cross slide of xingren (*Prunus armeniaca* L.) .

Instruments and appliances

Tweezers, magnifying glass, anatomical lens, dissector, microscope.

Contents and procedures

1. Simple fruit

1.1 Fleshy fruit Bringing the fruits of *Lycopersicon esculentum*, *Lycium chinensis*, *Citrus reticulata*, *Cucumis satious*, *Prunus davidiana* or *P. triloba* Lindl. , *Pyrus* Sp. , to crosscut and observe whether their outer, middle and inter rind are clearly to distinguished or not? What about their quality? What are the numbers of ovary and the type of placentation? How to distinguish between true or false fruit?

1.2 Dry fruit

1.2.1 Bringing the fruits of *Metaplexis japonica*, *Cassia troa*, *Glycine max*, *Capsella bursa -pastoris*, *Syringa oblata*, *Aristolochia contorta*, *Dioscotea nipponica*, *Dianthus chinensis* or *Agrostema githago*, *Datura stramonium*, to observe their form of dehiscent after matured, one seam dehiscent or two? Is its dehiscent among room or room - back or room - axile , distinguish their number of carpel and ovary and the type of placentation.

1.2.2 Observe the fruits of *Arctium lappa*, *Helianthus annuus*, *Zea mays*, *Castanea mollissima*, *Corylus heterophylla*, *Foeniculum vulgare* or *Archangelica dahurica*, *Eucommia ul-*

moides, *Acer negundo* , and to decide the type of indehiscent fruit according to the characters of the quality of rind, the difficult or easy of the dividing between rind and seed coat.

2. Aggregate fruit

Observe the fruits of *Paeonia lactiflora* or *P. suffruticosa*, *Illicum verum*, *Schisandra chinensis*, *Pulsatilla chinensis*, *Frgaria ananassa*, to understand their fruit to be composed of many divided carpels gathering on the receptacle, point out the type of aggregate fruit.

3. Collective fruit

Observe the fruits of *Morus alba* , *Aechmea weilbachii*, *Ficus carica*, to distinguish the structure of eating part, point out whether they are true or false fruit.

4. The structure of fruit

Observe the slide of cross section of *Lycium chinense* under microscope.

4.1 Exocarp It is composed of a line of flat cell. Cells wall are thinner and cuticle, and outer is small tooth - shaped raised.

4.2 Mesocarp It consists of lines of flat cells. 1 - 2 lines are smaller at the outer, and middle are bigger. Some cells have crystal sand of calcium oxalate. Vascular bundles are bicollateral bundle and have a dispersed arrangement.

4.3 Endocarp It composes of a line of ellipsed - shaped cell. Cells extend on the tangent.

4.4 Seed coat There is a line of similar rectangle stone cell at the outer, cells of side face and inside thicken to "U" . There are 3 - 4 lines crushed parenchyma cells under it. A layer of flat rectangle parenchyma cells lie in the most inter, and they are a little lignifications.

4.5 Parenchyma cells of endosperm, radicle and cotyledon They have fixed oil and pellet substance.

5. Albuminous seed and exalbuminous seed

5.1 Albuminous seed Observing the seed of *Ricinus communis*. It is flat and wide ovate, one surface is flatter and the other is thicker. The seed coat is hard, and are made up of 3 layers, outer layer is membranaceous with brown - black figure and lustre on it; the middle layer is bone, including brown - black pigment; the inter is white membraneceous. Observing the formation of the seed, and there is a light color sponged rise on the narrower end, named caruncle. It can inhale water, and benefits to sprouting of the seed. There is a rise in the caruncle of seed belly, named hilum, a north - south upper rise under caruncle and hilum, named raphe. The end of raphe is chalaza. Micropyle is covered by caruncle and can be not seen. Paying attention to strip seed coat, and the fat endosperm in it.

Cutting parallel to the wide surface of seed and endosperm being divided two parts, to observe under magnify glass, you can find two flat and clear - vein cotyledon, as well as radicle, very little plumule and short hypocotyls.

5.2 Exalbuminous seed Observing the seed of *Arachis hypogaea*: seed is wide ovate, seed coat is flat, pink or purplish red. Peel off seed coat, you can see two fat milky - white gloss cotyledons. Separate two cotyledon, and find its plumule is composed to a main bud and two lateral buds. Hypocotyl of *Arachis hypogaea* lie under plumule, thicker and cotyledon living on it. The lower of hypocotyl is radicle. It extends out two cotyledons and be short beak.

Moreover, bringing *Prunus armeniaca* to observe, and distinguish their formation and internal structure.

6. The structure of seed

Bringing the slide of cross section of *Prunus armeniaca* to observe.

6.1 Outer epidermis of seed coat It is composed of a line of flat cell, and dispersed rectangle and round yellow stone cells. A part of upper raises face, and below buries in the parenchyma.

6.2 Inter epidermis of seed coat It is composed of a line of flat cell, and has yellow something.

6.3 Perisperm It is composed of lines of decadent parenchyma.

6.4 Endosperm It is composed of a line of rectangle cell, and has aleurone grain and fixed oil.

6.5. Cotyledon It consists of lines of flat cell, and has aleurone grain. There is a small calcium oxalate crystal and fixed oil in the bigger aleurone grain.

Experiment report

1. List and compare to the type of fruit you have observed and distinguish between true or false fruit?

2. List and compare to their difference among follicle, legume, silique and capsule.

	follicle	legume	silique	capsule
the numbers of carpel				
the numbers of ovary				
the type of placenta				
the form of dehiscent				

3. Draw the structure of fruit of *Lycium chinense*, and note the name of every section

4. Draw the formation and section of seed of a *Ricinus communis*, and note the name of every part.

5. Draw the formation and section of seed of *Prunus armeniaca*, and note the name of every part.

Questions

1. How does normal fruit grow?

2. What is the basis of the class of fruit?

3. How does one distinguish between albuminous seed and exalbuminous seed.

4. How can you identify that embryo is young plant by experiment?

实验十 藻、菌、地衣植物

【目的要求】

1. 观察藻、菌、地衣等代表植物，了解各类植物的主要特征。

2. 识别常见药用藻、菌、地衣植物，熟悉低等植物的一般特征。

【实验材料】

标本：海带、蛋白核小球藻、石莼、甘紫菜、石花菜、海蒿子、麦角菌、黑木耳、茯苓、猪苓、灵芝、松萝、石耳。

【仪器用品】

显微镜、放大镜、解剖用具。

【内容与方法】

观察上述各类代表植物的标本，指出它们的主要特征，判断它们的植物体是孢子体还是配子体。

一、藻类

1. 褐藻类

（1）海带　植物体分三部分：呈假根状的固着器、柄、带片。为药材“昆布”的来源。横切面见附图 10 - 1.

（2）海蒿子　藻体直立，多分枝，分枝有叶状突起呈线形、披针形，气囊与生殖托生在小枝叶腋间。

2. 红藻门

（1）甘紫菜　藻体呈薄膜状，遇水后，手摸有黏滑感，紫红色或淡紫红色，全藻入药。

（2）石花菜　藻体扁平直立，丛生，紫红色或红棕色，羽状分枝 4 ~ 5 次，全藻入药。

3. 绿藻门

（1）石莼　藻体为膜状体，由两层细胞组成，基部具有多细胞固着器，全藻入药。

（2）蛋白核小球藻　植物体为单细胞，很小，卵圆形或圆球形，壁内有 1 个近似杯状的载色体和 1 个淀粉核，细胞之内有 1 个细胞核。

二、真菌类

1. 子囊菌亚门　麦角菌：常寄生在禾本科植物小穗上，呈圆柱状，红紫色，微弯曲如角，此为菌核，药用。

2. 担子菌亚门

（1）灵芝　子实体木栓质，菌柄着生部位位于菌盖一侧而非中间，菌盖半圆形至肾形，上面红褐色，有光泽，具环状横纹，下面（管孔面）白色，有许多小孔，内藏担孢子，子实体药用。见附图 10－2。

（2）猪苓　菌核常为不规则块状，表面凹凸不平、棕褐色至灰黑色，有光色，断面白色，菌核入药。

观察茯苓和黑木耳的外形。

三、地衣

观察松萝和石耳的外形。

【实验报告】

列表填写各类植物的主要特征。

	藻类	菌类	地衣
植物体			
营养方式			
色素			
世代交替			
生殖器官及生殖细胞			
合子发育方式			

【思考题】

低等植物有哪些共同特征？

Experiment 10　Algae, Fungi and Lichens Plant

Aim and demand

1. Observe algae, fungi, lichens and understanding their main characters.

2. Be capable to identify commonly used pharmaceutical algae, fungi, lichens. Gain common characteristics of elementary plants.

Experiment materials

Haidai (*Laminaria japonica* Aresch.); Danbaihexiaoqiuzao (*Chlorella pyrenoidosa* Chick.); Shichun (*Ulva lactuca* L.); Ganzicai (*Porphyra tenera* Kjellm.); Shihuacai (*Gelidium amansii* Lamx.); Haihaozi (*Sargassum pallidum* (Turn.) C. Ag.); Maijiaojun (*Claviceps purpurea* (Fr.) Tul.); Heimuer (*Auricularia auricula* (L. ex Hook.) Onder.); Fuling (*Poria cocos* (Schw.) Wolf.); Zhuling (*Polyporus umbellatus* (Per.) Fr.); Lingzhi (*Ganoderma lucidum* (Leyss. ex Fr.) Karst.); Songluo (*Usnea diffracta* Vain.); Shier (*Umbelicaria esculenta* (Miyoshi) Minks).

Instruments and appliances

Magnifying glass, dissector, microscope.

Contents and procedures

Observe simple kinds of delegate plants, point out their main characters, and decide if their plants are sporphyte or gametophyte.

1. Algae

1.1 Phaeophyta

1.1.1 *Laminaria japonica* The plant is divided three parts: the fixed organ of rhizome、stem and body band. It is the source of crude drugs Kunbu (*Ecklonia kurome* Okam.) (Figure 10 – 1).

1.1.2 *Sargassum pallidum* Algae is erect and have lots of branches, branches have leaf projectings, which shape is linear or lanceolate. Air sac and reproduction branch lie among leaf axil of short branches.

1.2. Rhodophyta

1.2.1 *Porphyra tenera* Algae is flat membranaceous, meeted water, feeling sticky and smooth, purple red or light purple red. Whole plant is medicinal materials.

1.2.2 *Gelidium amansii* Algae is flat, erect, growing thickly, purple red or red – brown, 4 – 5 pinnately branching, whole plant is medicinal materials.

1.3. Chlorophyta

1.3.1 *Ulva lactuca* Algae is flat membranaceous, and consists of 2 layers of cells. There is fixed organ of lots of cells at the conttom. Whole plant is medicinal materials.

1.3.2 *Chlorella pyrenoidosa* Algae is a small egg – round or round cell. There is a similiar cup – shaped chromoplast and starch, and a nucleus in the cell.

2. Eumycota

2.1. Ascomycotina

Claviceps purpurea Parasiting on the spike of graminease, cylinder, red purple, a little curved and similar horn. This is sclertium, and medicinal meterials.

2.2 Basidiomycotina

2.2.1 *Ganoderma lucidum* Sporophore is suberization, stipe lies aside of pileus doesn't the middle, pileus is from a half of orbicular to kidney, the upper is red – brown, glossy and rounded striate, the lower is white, a lot of little holes, and basidiospore is in it. Sclertium is medicinal materials (Figure 10 – 2).

2.2.2 *Polyporus umbellatus* Sclerotium is tuberous and irregular, the surface is concave – convex, brown, grey and black, glossy, the section is white. Sclerotium is medicinal materials.

Observing the formation of *Polyporus umbellatus*, *Poria cocos* and *Auricularia auricula*.

3. Lichens

Observing the formation of *Usnea diffracta* and *Umbelicaria esculenta*.

Experiment report

List the main characters of algae, fungi and lichens.

	algae	fungi	lichens
plant			
nutrition			
pigment			
alternation of generation			
organ and cell of reproduction			
growing form of zygote			

Questions

Write out the same characteristics in the elementary plants?

实验十一　苔藓植物、蕨类植物和裸子植物

【目的要求】

1. 观察苔藓植物和蕨类植物的代表植物，了解各类植物的主要特征。

2. 通过对苔藓植物、蕨类植物代表植物的观察，了解它们的生活史中两个不同的发育阶段，即有性世代和无性世代交替现象，加深对有性世代占优势及无性世代占优势两类植物世代交替的理解。

3. 掌握裸子植物的主要特征。

4. 识别常见的药用裸子植物。

【实验材料】

切片：地钱精子器托、地钱颈卵器托、葫芦藓雄配子体、土马骔雄配子体、蕨叶切片、蕨原叶体及幼孢子体。

卷柏、有柄石韦、粗茎鳞毛蕨。

油松的带雌球花、雄球花及球果的枝条。

油松雌球花、雄球花、银杏带种子枝条的浸泡标本。

侧柏、草麻黄的蜡叶标本。

【仪器用品】

显微镜、放大镜或解剖镜、解剖用具。

【内容与方法】

一、苔藓植物

葫芦藓：注意其植物是孢子体还是配子体，即有无根、茎、叶的分化，亦可用大金发藓（土马骔）作观察材料，孢蒴结构与葫芦藓基本相同。

在显微镜下观察地钱、葫芦藓、土马骔的石蜡切片，注意它们的颈卵器、精子器、孢蒴的特征。见附图 11－1～11－7。

二、蕨类植物

卷柏、有柄石韦、粗茎鳞毛蕨。

1. 观察上述3种蕨类植物的主要特征，主要观察内容如下。

（1）根状茎是横生、直生还是斜生。

（2）地上茎　多数种类没有，少数有地上茎，且不分枝或二叉分枝或轮生分枝。

（3）叶　分为小叶型、大叶型、同型叶（兼行光合作用与产生孢子囊）与异型叶（孢子叶、能育叶与营养叶、不育叶）。

（4）附属物　鳞片，毛。

（5）孢子囊穗或孢子囊群的有无，及在孢子叶上的着生位置（附图11－8）。

（6）孢子囊上着生的位置，顶生环带、横生环带、斜生环带、直生环带附图11－9）。

（7）孢子的形状，两面形、四面形、球状四面形、弹丝形（附图11－10）。

2. 在显微镜下观察蕨叶、蕨类原叶体及幼孢子体的石蜡切片，注意它们的孢子囊、原叶体及幼孢子体的特征。

三、裸子植物

（1）取油松的雄球花（小孢子叶球）在放大镜或解剖镜下观察外形，用镊子取一个雄蕊于载玻片上，置放大镜下观察其花粉囊（小孢子囊）的形状。

（2）取油松的雌球花（大孢子叶球），用放大镜观察外形，是由多数螺旋状排列的珠鳞（心皮、大孢子叶）组成。用刀片将雌球花纵切，注意珠鳞排列情况。剥开一片完整的珠鳞可见到腹面基部着生2枚胚珠，背面基部托生一小片苞鳞，与珠鳞分离。

（3）取侧柏和草麻黄的标本，观察其形状特征及雄球花、雌球花的外部形态。

（4）观察银杏标本，有长短枝之分，观察叶片的形状及脉序的类型，种子裸生，注意三层种皮的特征。

【实验报告】

1. 绘蕨类植物的生活史简图，注明有性世代和无性世代。

2. 列表填写苔藓、蕨类植物的主要特征。

	苔藓	蕨类
植物体		
营养方式		
世代交替		
生殖器官及生殖细胞		
合子发育方式		

3. 绘油松雌球花纵切面简图，并注明各部分名称。

4. 列表记录观察的裸子植物的名称、科名、药用部分及主要形态特征。

【思考题】

1. 苔藓植物的主要特征是什么？在生活史中哪个世代占优势？
2. 蕨类植物的主要特征是什么？它比苔藓植物进化表现在哪些方面？
3. 蕨类植物的世代交替如何？它的哪一世代最为发达？通常生活在什么地方？
4. 裸子植物的主要特征有哪些？为什么这些特征比蕨类植物适应陆生环境更有利？

Experiment 11 Bryophyta, Pteridophyta and Gymnosperma

Aim and demand

1. Understand main characteristics of bryophyta and pteridophyta through observe the delegate plants of the two types of plants.

2. Observe delegate plants of bryophyta and pteridophyta, to understand their living history in two different reprodution, the alternation phenomenon of sexual generation and asexual generation, deepen comprehension to two plants having different advantage in sexual and asexual generation.

3. Identify the main characteristics of gymnospermae.

4. Be capable to identify commonly used medicinal gymnospermae plants.

Experiment materials

Sections: antheridium and archegonium branch of Diqian (*Marchantia polymorpha* L.) male gametophyta of Huluxian (*Funaria hygrometrica* Hedw.), male gametophyta of Dajinfaxian (*Polytrichum commune* L.), the section slide of pteridophyta leaf. The prothallus and young sporophyte of pteridophyta; Juanbai (*Selaginella tamariscina* (Beauv.) Spring.); Youbingshiwei (*Pyrosia petiolosa* (Christ) Ching.); Cujinglinmaojue (*Dryopteris crassirhizoma* Nakai.).

Branches of Yousong (*Pinus tabulaeformis* Carr.) with female cone, staminate strobilus and cone.

Immersioning sample of Yinxing (*Ginkgo biloba* L.); wax sample of Cebai (*Platycladus orientalis* (L.) Franco.) and Caomahuang (*Ephedra sininca* Stapf.).

Instruments and aplliances

Microscope, Magnifying glass or anatomical lens, dissector.

Contents and procedures

1. Bryophyta

Funaria hygrometrica: Observe its plant, decide whether it is sporophyta or gametophyte, whether it has the differentiation of root, stem, leaf. You can also observe *Polytrichum commune*, it's structure of capsule is the same as *Funaria hygrometrica*.

Observe the wax slide of *Marchantia polymorpha*, *Funaria hygrometrica*, *Polytrichum commune* under the microscope, pay attention to their characters of archegonium, antheridium and capsule. (Figure 11-1 to 11-7).

2. Pteridophyta

2.1 *Selaginella tamariscina*, *Pyrosia petiolosa*, *Dryopteris crassirhizoma*: Observe the main characters of three pteridophyta above, the main contents are as follows.

2.1.1 Rhizome is orthotropous, hemitropous or campylotropous.

2.1.2 Ground stem Most of the pteridophyta plants don't have ground stem, few ground stem are not branch or dichotomous or verticillate branching.

2.1.3 Leaf Leaves included microphyll, macrophyll, homomorphic leaf (the same photosynthesis and producing sporangium) and heteromorphic leaf (sporophyll, fertile frond, foliage leaf and sterile frond).

2.1.4 Subsidiary accessory, scale hair.

2.1.5 Whether exist sporophyll spike or sorus, and location on the sporophyll (Figure 11-8).

2.1.6 Annulus locates on the sporangium, including apical orthotropous, hemitropous, campylotropous annulus (see figure 11-9).

2.1.7 The shape of spore: two aspects, four aspects, rounded four aspects, spring (Figure 11-10)

2.2 Observe the wax slide of leaf of pteridophyta, prothallus and little sporophyte under the microscope, pay attention to their characters of sporangium, prothallus and young sporophyte.

3. Gymnospermae

3.1 Bringing staminate strobilus of *Pinus tabulae* formis (small strobilus) to observe under magnifying glass or anatomical lens, and bring a stamen with tweezers on the glass, observe its formation of pollen sac (small sporangium) under magnifying glass.

3.2 Bring female cone (big strobilus) of *Pinus tabulaeformis*, to observe the formation. It is compose of lots of spiral arrange pearl scale (carpel, big sporophyll). Cutting female cone with knife, observing arrange situation of pearl scale.

Stripping a whole pearl scale, you can see 2 existing ovules on the belly basis, and lie a bract scale on back basis, separated with pearl scale.

3.3 Bring sample of *Platycladus orientalis* and *Ephedra sininca*, to observe the morphological characteristics and formation of staminate strobilus and female cone.

3.4 Observing sample of *Ginkgo biloba*, you can see long shoot and short shoot, pay attention to the shape of leaves and definite its' type of venation. The seed is exposed, and it has three layer of seed coat.

Experiment report

1. Draw the living history of pteridophyta, and note sexual and asexual generation.
2. List the main characteristics of bryophyta and pteridophyta.

	bryophyta	pteridophyta
plant		
nutrition		
alternation of generation		
organ and cell of reproduction		
growing form of zygote		

3. Draw vertical section of the female cone of *Pinus tabulaeformis*, note the name of every parts.

4. List familia name, medicinal parts and main characteristics of gymnospermae plants you have observed in this experiment.

Questions

1. What are the main characteristics of bryophyte? Which generation is advantageous in living history?

2. What are the main characteristics of pteridophyta? Which aspects are shown in the evolution thanbryophyte?

3. Describe the alternation of generation of pteridophyta? Which generation is the most developed? What place is it usually living in?

4. List the characteristics of a gymnospermae plant, Why is the gymnospermae plant with these characteristics more adapted to land environment than the pteridophyta?

实验十二　桑科、马兜铃科、蓼科、毛茛科

【目的要求】

1. 掌握桑科、马兜铃科、蓼科、毛茛科植物的主要形态特征。
2. 识别各科的主要药用植物。

【实验材料】

一、桑科

1. 代表植物　桑（*Morus alba* L.）枝条、花序及果序标本。

2. 其他药用植物标本　大麻（*Cannabis sativa* L.）蜡叶标本。

二、马兜铃科

1. 代表植物　北细辛（*Asarum heherotropoides* Fr. Schmidt var. *mandshurium*（Maxim.）Kitag.）植株及花果。

2. 其他药用植物标本　华细辛（*A. sieboldii* Miq.）蜡叶标本；汉城细辛（*A. sieboldii* Miq. var. *seoulense* Nakai）蜡叶标本。

三、蓼科

1. 代表植物 何首乌（*Polygonum multiflorum* Thunb.）植株及花果。

2. 其他药用植物标本 药用大黄（*Rheum officinalis* Bill.）植株及花果；掌叶大黄（*R. palmatum* L.）植株及花果；唐古特大黄（*R. tanguticum* Maxim. ex Balf.）植株及花果；波叶组大黄：华北大黄（*R. franzenbachii* Münt.）或藏边大黄（*R. emodi* Wall）植株及花果；虎杖（*Polygonum cuspidatum* Sieb. et Zucc.）植株及花果。

四、毛茛科

1. 代表植物 北乌头（*Aconitum kusnezoffii* Reichb.）的植株、花果。

2. 其他药用植物 黄连（味连）（*Coptis chinensis* Franch.）植株标本；三角叶黄连（雅连）（*C. deltoidea* C. Y. Cheng et Hsiao）植株标本；云南黄连（云连）（*C. teeta* Wall.）植株标本；升麻（*Cimicifuga foetida* L.）带花植株。

【仪器用品】

放大镜、解剖镜、解剖用具。

【内容与方法】

一、解剖下列各代表植物

1. 桑科 桑：取枝条观察，树皮黄褐色，单叶互生，卵形或宽卵形；葇荑花序；花单生，雌雄异株，分别从雌株和雄株的花序上各取一朵小花观察，可见雄花花被片4，对瓣生长；雄花花被片4。无花柱，柱头2裂，子房上位。瘦果包于肉质花被片内，密集成聚花果。果实（桑葚）及根皮（桑白皮）入药。

2. 马兜铃科 北细辛（附图12－1）：草本，根状茎横生，多生细长的须根，有辛香味。叶基生，有长柄，叶片卵状心形或近肾形，全缘。花单生，注意花梗在近花被管处弯曲，花被管壶形或半球形，顶端三裂，裂片向外反折。雄蕊12，着生于子房中下部，花丝与花药等长；子房半下位，花柱6，顶端2裂，蒴果，半球形。全草入药。

3. 蓼科 何首乌：缠绕草本，块根表面红褐色。叶互生，卵状心形，具膜质托叶鞘，抱茎。圆锥花序，花白色，较小，花被2裂，外侧3片，背部有翅。瘦果具3棱。块根及藤茎入药。

4. 毛茛科 北乌头（附图12－2）：取植株观察，注意块根的形状，叶片3全裂，总状花序，萼片5，蓝紫色，上萼片盔状，花瓣2，最下一片可见有长爪，雄蕊多数，心皮3～5离生。观察蜡叶标本可见果实为聚合蓇葖果。块根入药称为乌头。

二、其他药用植物标本观察

1. 桑科 大麻：草本，叶互生，掌状全裂，裂片披针形。花单性异株，雄株上可见雄花排成圆锥花序，雌花丛生于叶腋，每朵花有一卵形苞片，花被片1，膜质；瘦果。果实（火麻仁）入药。雌株幼嫩果穗可制成毒品。

2. 马兜铃科

（1）华细辛　与北细辛的区别在于根茎较长，节间距离均匀，叶端渐尖，花被裂片直立，不反折。全草入药。

（2）汉城细辛（附图 12－3）　本品为华细辛变种，区别为叶柄有毛，叶下面通常密生较长的毛。全草入药。

3. 蓼科

（1）药用大黄　草本，根及根茎肥厚，折断后，断面黄色有星点；叶片圆形，掌状浅裂，浅裂片呈大齿形或宽三角形，托叶鞘筒状。花较大。瘦果具 3 棱，沿棱生翅。根及根茎入药。

（2）掌叶大黄　与药用大黄的主要区别为叶片宽卵形或近圆形，掌状半裂，裂片再羽裂，花梗纤细，中下部有关节。根及根茎入药。

（3）唐古特大黄　与掌叶大黄相似，本种叶片深裂，裂片通常窄长，呈三角状披针形或窄条形。根及根茎入药。

（4）波叶组大黄　观察华北大黄和藏边大黄的植株，与上述掌叶组 3 种大黄的主要区别为叶缘具不同程度的皱波，叶片完整不裂。根及根茎入药，称土大黄。

（5）虎杖（附图 12－4）　高大草本，根状茎粗大，茎上生红色或紫红色斑点，节间明显，叶片阔卵形。根状茎入药。

4. 毛茛科　观察 3 种黄连的植物标本，注意它们的形态特征。

（1）黄连　根状茎黄色，分枝成簇状，叶 3 全裂，中央裂片具细柄，侧生裂片不等 2 深裂，边缘有尖锐锯齿，根茎入药。

（2）三角叶黄连　根状茎不分枝，叶的一回裂片的深裂片彼此邻接，根茎入药，称为雅连。

（3）云南黄连　根状茎分枝少而细，叶的羽状深裂片彼此疏离。花瓣匙形，先端钝圆。根茎入药，称为云莲。

（4）升麻　根状茎粗壮，具有多数内陷的圆洞状老茎残迹。观察其复叶是几回羽状复叶，小叶片卵形或菱形。圆锥花序顶生。聚合蓇葖果，根状茎入药。

【实验报告】

1. 写出桑科植物的主要特征。
2. 写出蓼科植物的主要特征。
3. 比较正品大黄药材的几种植物来源的主要异同点。
4. 写出马兜铃科植物的主要特征。
5. 列出毛茛科植物的主要特征。

【思考题】

1. 蓼属、酸模属、大黄属有何区别？
2. 比较毛茛科中乌头属、毛茛属、铁线莲属的主要异同点，并各列出 2 种药用植物。

Experiment 12 Mulberry Family (Moraceae), Birthwort Family (Aristolochiaceae), Buckwheat Family (Polygonaceae), Crowfoot Family (Ranunculaceae)

Aim and demand

1. Grasp the main morphological characters of Mulberry family (Moraceae), Birthwort family (Aristolochiaceace), Buckwheat family (Polygonaceae) and Crowfoot family (Ranunculaceae).

2. Recognize the main medicinal plants of Mulberry family (Moraceae), Birthwort family (Aristolochiaceace), Buckwheat family (Polygonaceae) and Crowfoot family (Ranunculaceae).

Experiment materials

1. Mulberry family (Moraceae)

1.1 Typical plant Sang (*Morus alba* L.) with branches, inflorescences.

1.2 Other samples of pharmaceutical plants The herbarium sheets of Dama (*Cannabis sativa* L.).

2. Birthwort family (Aristolochiaceace)

2.1 Typical plant The total herbs of Beixixin (*Asarum heterotropoides* Fr. Schmidt var. *mand – shurium* (Maxim.) Kitag.).

2.2 Other samples of pharmaceutical plants The herbarium sheets of Huaxixin (*A. sieboldii* Miq.) and Hanchengxixin (*A. sieboldii* Miq. var. *seoulense* Nakai).

3. Buckwheat family (Polygonaceae)

3.1 Typical plant The total herbs of tuber of Heshouwu (*Polygonum multiflorum* Thunb.) with flowers and fruits.

3.2 Other samples of pharmaceutical plants The total herbs with flowers and fruits of Yaoyong – dahuang (*Rheum officinalis* Bill.), Zhangyedahuang (*R. palmatum* L.), Tangutedahuang (*R. Tanguticum* Maxim. ex Balf.), Huabeidahuang (*R. franzenbachii* Miint.) and Zangbiandahuang (*R. emodi* Wall) within the group of undulate rhubarb as well as the herbs of Huzhang (*Polygonum cuspidatum* Sieb. et Zucc.).

4. Crowfoot family (Ranunculaceae)

4.1 Typical plant The total herb with flowers and fruit of Beiwutou (*Aconitum kusnezoffii* Reichb.).

4.2 Other samples of medicinal plants Huanglian (*Coptis chinensis* Franch.), Sanjiaoyehuanglian (*C. deltoidea* C. Y. Cheng et Hsiao), Yunnanhuanglian (*C. teeta* Wall.), Shengma (*Cimicifuga foetida* L.).

Instruments and apliances

Magnifying glass, anatomical lens, Dissector.

Contents and procedures

1. Observe and dissect the typical plants as follows

1.1 Mulberry family (Moraceae)

Morus alba : Take branches for an observation, you may find such characters as tawny bark, alternating single leaves which are ovate or a bit broad. The unisexual flowers are arranged as catkin. In fact, they are dioecim. Observing flowers from male plants and female plants respectively, you may see 4 perianths and 4 stamens, growing oppositely in male flowers with 4 perianths, no style, bidivided stigma, superior ovary in female plants. There are some achenes sphered by succulent perianth forming into collective fruits. Fruits and bark in drug use.

1.2 Birthwort family (Aristolochiaceace)

Asarum heterotropoides var. *mandshurium* (Figure 12 – 1): As a kind of herb, its rhizomes grow horizontally with many tenuous but long fibrous roots and spicy tinge, its basal leaves which are cordate or reniform and entire have long petioles, and its flowers grow singly. Please pay attention to the turns of pedicels to perianth tubes which are semiorbicular or like a pot with the tops tri – dividing and then the separated pieces inversely tucking. There are 12 stamens in the middle or inferior ovary, filaments and anthers with equal length. Besides, other features like half – inferior ovary, 6 bidivided styles, semiorbicular capsules also can be seen. Total herbs is used in making drugs.

1.3 Buckwheat family (Polygonaceae)

Polygonum multiflorum : Twining vine. The tuberous root is reddish – brown. As far as the morphology is concerned, the anplexicaul leaves which are ovate or cordate grow alternately, owning membranaceous ocreas; the white small flowers have their perianths divided into 5 pieces of which 3 have spurs in the back, also the achenes have 3 arrises. Tuberous root and vines in drug use.

1.4 Crowfoot family (Ranunculaceae)

Aconitum kusnezoffii (see figure 12 – 2): Observe the whole plant, pay attention to the shape of root tuber; leaf 3 divided; indefinite inflorescence; iodes sepals 5, blue and purple, the up one helmet – shaped; 2 petals, with long claw; stamens numerous; free carpels 3 ~ 5; fruit is polysecus; root tuber in drug use called "Wu – Tou" .

2. Observe other samples of pharmaceutical plants

2.1 Mulberry family (Moraceae)

Cannabis sativa: A kind of herb with its lanceolate leaves growing alternately, palmately divided. As dioecim, on the male plants we can see male flowers form into panicle while female flowers which have a ovate bract and membranaceous preanth in each fasioled in axils on the female plants. It's seeds are used in making drugs and tender fruit spikes of male plants are narcotic drug.

2.2 Birthwort family (Aristolochiaceace)

2.2.1 *Asarum sieboldii* Here are some differences from *A. heterotropoides*: longer rhizomes, even internodes, acuminate leaves - sharp, divided, vertical and non - tucking perianth pieces. Total herbs is used in making drugs.

2.2.2 *A. sieboldii* var. *seoulense* (Figure 12 - 3) A variate of Chinese asarum, but there is fluff on the petioles and the back of the leaves (always much longer). Total herbs is used in making drugs.

2.3 Buckwheat family (Polygonaceae)

2.3.1 *Rheum officinalis* Herb. Its roots and rhizomes whose cutting sections are yellow with many spots are fleshy. The leaves are round, palmately lobate and the pieces are dentate or broad - triangled. It also has tubular ocreas and relatively bigger flowers. The achene has 3 arrises where septum rise. Root and rhizomes are used in making drugs.

2.3.2 *R. palmatum* The main differences from *R. officinalis* are that its leaves are broad - ovate or like round, palmately parted at first and then pinnately separated, its pediceles are more tenuous and have nodes in the middle or lower parts. Roots and rhizomes are used in making drugs.

2.3.3 *R. Tanguticum* It is opposite from *R. palmatum* for its leaves often deeply divide into narrow and long pieces which are triangled - lanceolate oe linear. Roots and rhizomes in drug use.

2.3.4 *R. franzenbachii* and *R. emodi* are belong to the group of undulate rhubarb. Being different from the 3 rhubarbs mentioned above, these two plants have their leaves undulate in different degrees and divided totally. Roots and rhizomes in drug use.

2.3.5 *Polygonum cuspidatum* (Figure 12 - 4) High herb. The rhizomes are strong and the stems often with some red or magenta spots have obvious internodes, leaves are broad - ovate. Rhizomes are used in making drugs.

2.4 Orowfoot family (Ranunculaceae)

2.4.1 *Coptis chinensis* yellow rhizome; fascioled branches; leaves 3 divided, the middle lobate has slender stipe, while the lateral lobate has unequal 2 parted with sharp serrate on the edges; rhizome is used in making drugs.

2.4.2 *Coptis deltoidea* rhizome with no branches; leaf segment of the first division is contiguous each other; rhizome called YaLian.

2.4.3 *Coptis teeta* rhizome only has little branches while slender; feather shape segment of the leaves are separate from each other. Petals like spatulate, blunt edge part; rhizome is used in making drugs, called YunLian.

2.4.4 *Cimicifuga foetida* stout rhizome with many depressed vesitages of stems; the type of pinnate compound leaf is bipinnate or tripinnate. The leaflet is ovate or rhombic; terminal panicle; fruit is polysecus; rhizome is used in making drugs.

Experimentreports

1. Write down the main characters of mulberry family (Moraceae).

2. Write down the main characters of Buckwheat family (Polygonaceae).

3. Show the parallels and differences between the three certified rhubarbs with different origins.

4. Show the main characters of Birthwort family (Aristolochiaceace).

5. List the main characteristics of crowfoot family (Ranunculaceae).

Questions

1. What are the differences between Polygonum, Rumex and Rheum?

2. Describe the main difference and similarities among Ranunculus, Aconitum and Clematis of crowfoot family (Ranunculaceae), and list two pharmaceutical plants of every genus.

实验十三　木兰科、芍药科、罂粟科、十字花科

【目的要求】

1. 掌握木兰科、芍药科、罂粟科、十字花科植物的主要形态特征。
2. 识别各科主要的药用植物。

【实验材料】

一、木兰科

1. 代表植物　北五味子（*Schisandra chinensis*（Turcz.）Baill.）的标本、花、果实。

2. 其他药用植物标本　华中五味子（*S. sphenanthera* Rehd. et Wils.）标本、花、果实；南五味子（*Kadsura longipedunculata* Finet. et Gagnep.）标本、花、果实；厚朴（*Magnolia officinalis* Rehd. et Wils.）枝条、花。

二、芍药科

1. 代表植物　芍药（*Paeonia lactiflora* Pall.）的植株、花、果。

2. 其他药用植物　牡丹（*Paeonia suffruticosa* Andr.）植株、花、果。

三、罂粟科

1. 代表植物　白屈菜（*Chelidonium majus* L.）带花全株。

2. 其他药用植物标本　延胡索（*Corydalis yanhusuo* W. T. Wang）植株、罂粟（*Parpaver somniferm* L.）带花植株。

四、十字花科

1. 代表植物　大白菜（*Brassica pekinensis*（Lour.）Rupr.）带花全株、果。

2. 其他药用植物标本　菘蓝（*Isatis indigotica* Fort.）、白芥（*Sinapis alba* L.）带花

全株、果。

【仪器用品】

放大镜、解剖镜、解剖用具。

【内容与方法】

一、观察解剖下列各代表植物

1. 木兰科　五味子（附图 13－1）：木质藤本，叶片阔椭圆形或倒卵形，边缘具腺齿，花单性异株，新鲜的花被片乳白色至粉红色，摘除花被片后，再逐个摘除雄蕊和离生的雌蕊，记录各自的数目。取浸泡的果实标本可见聚合浆果排在延长的花托上成穗状。果实入药。

2. 芍药科　芍药（附图 13－2）：草本。根粗壮，取带叶的花枝观察，下部叶多二回三出复叶，小叶窄卵形，或窄椭圆形，花大而艳丽；花萼，花瓣离生；雄蕊多数；心皮 2～5 离生。注意观察花盘肉质，仅包裹心皮基部。蓇葖果卵形。根入药。

3. 罂粟科　白屈菜（附图 13－3）：取带花全株观察。鲜材料的茎叶折断后有黄色乳汁流出，叶互生，羽状全裂，花瓣4，黄色；观察2枚萼片花开后脱落与否，横切或纵切子房，注意心皮数、子房室数、胚珠的数目及胎座类型。观察果实类型。全草入药。

4. 十字花科　大白菜：取带花植株观察叶序、叶片及花序的类型后，摘取一朵花观察。注意花冠的类型、萼片及花瓣的数目，雄蕊的个数，是否为四强雄蕊，有无蜜腺及蜜腺数目，用放大镜观察心皮数，假隔膜及子房室数，判断胎座类型。取近成熟的果实纵剖观察种子的着生位置并判断果实类型。

二、其他药用植物标本的观察

1. 木兰科

（1）华中五味子　与北五味子相似，仔细观察标本上的花被片 8 枚，鲜时橙黄色，计数雄蕊及心皮的数目。聚合浆果，与北五味子比较，果肉较薄。

（2）南五味子　木质藤本；叶近革质，椭圆形，叶缘可见稀疏锯齿；花单性异株，花托果时不延长；聚合浆果。果实，根皮及茎叶入药。

（3）厚朴　乔木，叶大，革质，倒卵形，集生枝顶；花顶生，大，白色，厚肉质；聚合蓇葖果，茎皮、根皮及花药用。

2. 芍药科　牡丹（附图 13－4）：灌木，常二回三出复叶，顶生小叶 3 裂，侧生小叶不等 2 浅裂，花单生枝顶，萼片 5，宿存；花瓣 5 或为重瓣；花盘杯状，包住心皮。蓇葖果，密生褐黄色毛。根皮入药。

3. 罂粟科

（1）延胡索（附图 13－5）　草本，块茎扁球形，叶二回三出全裂，末回裂片披针形，总状花序顶生，花瓣 4，有一瓣基部有长距，块茎药用。

（2）罂粟　全株粉绿色，折断后有白色乳汁，叶互生，基部抱茎，花大型单生，萼片 2，早落，子房由多心皮组成，1 室，侧膜胎座。取果实观察其形状、大小，判断果实类型。未成熟果实取乳汁，可制成鸦片，果壳亦作药用。

4. 十字花科

（1）菘蓝（附图 13－6）　主根圆柱形，茎上部多分支，叶互生，茎生叶稍带粉霜，基部半抱茎，短角果长圆形扁平，边缘有翅。根称为板蓝根。叶作大青叶入药。

（2）白芥　全株被白色粗毛，茎基部的叶具长柄，长角果圆柱形，密被白色柔毛，先端具长喙。种子入药。

【实验报告】

1. 绘厚朴花的解剖图，并注明各部分名称。
2. 比较芍药科与毛茛科的异同点。
3. 列出罂粟科的主要特征，并列举该科数种常见药用植物及药用部位。
4. 说明十字花科的特征，并比较角果、蒴果、蓇葖果的异同点。

【思考题】

1. 为什么说木兰科植物是现存的植物中的原始类群？
2. 比较毛茛科与木兰科的主要异同点。
3. 罂粟科药用植物的主要药用部位有哪些？有哪些生物活性？
4. 中药大青叶的基源植物共有哪几种？主要区别点是什么？

Experiment 13　Magnolia Family（Magnoliaceae），Peony Family（Paeoniaceae），Poppy Family（Papaveraceae），Mustard Family（Cruciferae）

Aim and demand

1. Grasp the main morphological characteristics of magnolia family（Magnoliaceae），peony family（Paeoniaceae），poppy family（Papaveraceae）and mustard family（Cruciferae）.

2. Be capable to identify the main pharmaceutical plants of magnolia family（Magnoliaceae），peony family（Paeoniaceae），poppy family（Papaveraceae）and mustard family（Cruciferae）.

Experiment materials

1. Magnolia family（Magnoliaceae）

1.1 Typical plant　The total herbs，flower，fruit of Beiwuweizi（*Schisandra chinensis*（Turcz.）Baill.）.

1.2 Other samples of medicinal plants　Huazhongwuweizi（*S. sphenanthera* Rehd. et Wils.），Nanwuweizi（*Kadsura longipedunculata* Finet. et. Gagnep.），Houpu（*Magnolia officinalis* Rehd. et Wils.）.

2. Peony family（Paeoniaceae）

2.1 Typical plant　The total herbs，flower，fruit of Shaoyao（*Paeonia lactiflora* Pall.）.

2.2 Other samples of medicinal plants Mudan (*Paeonia suffruticosa* Andr.).

3. Poppy family (Papaveraceae)

3.1 Typical plant The total herbs of Baiqucai (*Chelidonium majus* L.) with flowers.

3.2 Other samples of medicinal plants Yingsu (*Papaver somniferum* L.) and Yanhusuo (*Corydalis yanhusuo* W. T. Wang).

4. Mustard family (Cruciferae)

4.1 Typical plant The total herbs of Baicai (*Brassica pekinensis* (Lour.) Rupr.) with flowers and fruits.

4.2 Other samples of medicinal plants Total herbs of Songlan (*Isatis indigotica* Fort.), Baijie (*Sinapis alba* L.).

Instruments and aplliances

Magnifying glass, anatomical lens, dissector.

Contents and procedures

1. Observe and dissect the typical plants as follows

1.1 Magnolia family (Magnoliaceae)

Schisandra chinensis (Figure13 – 1): woody vine; elliptical or obovate leaf, with glandulosa – serratus on the edges; dioecism flower; color of fresh perianth from lacte to pink; count the number of stamens and free pistils; observe the soak sample of fruit, berries many, on an elongated receptacle like spicate; root is used in making drugs.

1.2 Peony family (Paeoniaceae)

Paeonia lactiflora (Fisure 13 – 2): herb; stout root; the lower leaf is biternate compound leaf, leaflet ovate or elliptical; showy flower; petal and sepal are distinct; stamen numerous; free carpels 2 – 5; succulent disc, only enclosing the base of carpels; ovate follicle; root is used in making drugs.

1.3 Poppy family (Papaveraceae)

Chelidonium majus (see figure13 – 3): observe the total herbs with flower. Fresh stem and leaf has yellow juice; leaf alternate, pinnatisect, with white powder; petals 4, yellow; you should know if 2 sepals fall down after the flower opens; cross cut or vertically cut ovary, count the number of carpels, ovary cells, ovules and which type of placenta it is. Observe the type of fruit. Total herb is used in making drugs.

1.4 Mustard family (Cruciferae)

Brassica pekinensis: observe the total herbs with flower. Get the type of phyllotaxy, blade and inflorescence; pay attention to the type of corolla and the number of petals, sepals and stamens, to decide whether the stamen is tetradynamous stamen or not; whether there is honey cup or not, and if there are, to count the number. By means of a magnifier to count the numbers of carpels, septulates and ovary cells, then to decide the type of placenta. Dissect premature fruit vertically, observe where the seed grows in the fruit and determine the type of fruit.

2. Observe other samples of pharmaceutical plants

2. 1 Magnolia family (Magnoliaceae)

2. 1. 1 *Schisandra sphonanthera* similar to chinese magnoliavine; tepals 8, orange; count the numbers of stamens and carpels; berry; compared with chinese magnoliavine, it has thinner pulp.

2. 1. 2 *Kadsura longipedunculata* woody vine; leaf is near coriaceous, elliptical, with little serrate on the edges; dioecism flower; receptacle not prolonged to berry while making a fruit; fruit, root – bark, stem and leaf are used in making drugs.

2. 1. 3 *Magnolia officinalis* tree; large leaf assembles on the top of branches, coriaceous, obovate; large white flower, terminal and succulent; polysecus; bark, root – bark and flower are used in making drugs.

2. 2 Peony family (Paeoniaceae)

Paeonia suffruticosa (Figure13 – 4): shrub; leaf usually is biternate compound leaf, the terminal lobe is 3 divided, and the lateral lobe is unequal 2 parted; simple flower on the top of branches; persistent calyx, 5; petals 5 or double flower; flower disc like cup, enclosing the carpels; follicle with thick testaceous hairs; root – bark is used in making drugs.

2. 3 Poppy family (Papaveraceae)

2. 3. 1 *Corydalis yanhusuo* (Figure13 – 5) herb, tube spherical, flat; bipinnate – ternately compound leaf, divided, top leaflet is lancedate; terminal raceme; petal 4, one of them has a long spur at its base. The tube is used in making drugs.

2. 3. 1 *Papaver somniferum* the total herb is somewhat green; can give out white milky juice when it be snaped; leaves alternate, amplexicaul at leaf base; flower showy, solitary; sepals 2, fall off early; ovary, many carpels united into 1 celled, parietal placenta; observe the shape and size of the fruit, then to determine the type of fruit; the juice of premature fruit can be made into opium. The fruit shell is used in making drugs.

2. 4 Mustard family (Cruciferae)

2. 4. 1 *Isatis indigotica* (Figure 13 – 6) the main root is terete; at the top of stem there are many branches; leaves alternate; cauline leaf with somewhat white powder, at its bottom half – clasping; silicle orbicular, flat, and the edge likes wing; its root is dyers woad root, its leaf is indigo – coloured woad leaf in traditional Chinese medicine.

2. 4. 2 *Sinapis alba* the total plant has white thick hair; the leaf at the bottom of stem has long petiole; silique terete dense with white villi, apex of fruit likes beak. The seed is used in making drugs.

Experiment report

1. Draw transverse section of official magnolia immature flower and note the name of each part.

2. Describe the main difference and similarities between crowfoot family (Ranunculaceae) and peony family (Paeoniaceae).

3. List the main characteristics of poppy family (Papaveraceae) and several commonly used pharmaceutical plants and pharmaceutical section.

4. List the main characteristics of mustard family (Cruciferae), and compare capsule, silicle, silique with follicle, noting the similarities and the differences.

Questions

1. Why plants of Magnolia family (Magnoliaceae) are original type among present existing plants?

2. Compare crowfoot family (Ranunculaceae) with magnolia family (Magnoliaceae) and list the main differences and similarities.

3. Describe the main pharmaceutical plants and pharmaceutical section of Poppy family (Papaveraceae), and list the main biological activity.

4. List the original plants of Traditional Chinese Medicine indigo - coloured woad leaf (Daqingye) and the differences between them.

实验十四　蔷薇科、豆科、五加科

【目的要求】

1. 掌握蔷薇科、豆科、五加科植物的主要形态特征。
2. 识别各科主要的药用植物。

【实验材料】

一、蔷薇科

1. 绣线菊亚科　绣线菊（*Spiraea salicifolia* L.）的枝条、花、果。

2. 蔷薇亚科　龙牙草（*Agrimonia pilosa* Ledeb.）带花全草。

3. 梅亚科　杏（*Prunus armeniaca* L.）的枝条、花、果。

4. 梨亚科　山楂（*Crataegus pinnatifida* Bge.）的枝条、花、果。

5. 其他药用植物　地榆（*Sangusorba officinalis* L.）带花植株；贴梗海棠（*Chaenomeles speciosa*（Sweet）Nakai）带花枝条一段及果实。

二、豆科

1. 含羞草亚科　代表植物：合欢（*Albizia julibrissin* Durazz.）的枝条、花、果。

2. 云实亚科

（1）代表植物　决明（*Cassia tora* L.）的带果枝条。

（2）其他药用植物　皂荚（*Gleditsia sinensis* Lam.），望江南（*Cassia occidentalis* L.）带果枝条。

3. 蝶形花亚科

（1）代表植物　甘草（*Glycyrrhiza uralensis* Fisch.）全株、花、果。

（2）其他药用植物　膜荚黄芪（*Astagalus membranaceus*（Fisch.）Bunge.），苦参

（*Sophora flavescens* Ait.）。

三、五加科

1. 代表植物 人参（*Panax ginseng* C. A. Mey.）全株带花、果。

2. 其他药用植物标本 西洋参（*Panax quinquefolium* L.），刺五加（*Acanthopanax senticosus*（Rupr. Et. Maxim.）Harms），三七（*Panax notoginseng*（Burk.）F. H. Chen）。

【仪器用品】

放大镜、解剖镜、解剖用具。

【内容与方法】

一、观察解剖下列各代表植物

1. 蔷薇科

（1）绣线菊 取枝条观察。注意叶序、叶形，观察有无托叶，属何种花序类型。取一朵小花放在放大镜下观察萼片、花瓣、雄蕊、雌蕊的数目。注意花筒的形状，是周围花还是下位花，观察果实类型。

（2）龙牙草（附图 14－1） 取带花全株观察。密被长柔毛，奇数羽状复叶，小叶 5～8 片，大小不等相间。注意小叶的形状及大小的不同。顶生总状花序。取花观察。花筒顶端有无一圈钩状刚毛，横切子房观察心皮数目，并观察果实类型。全草及根芽入药。

（3）杏 取枝条观察，注意叶形，观察有无托叶，取一朵小花观察萼片、花瓣、雄蕊的数目。纵剖花，注意花筒的形状，花被及雄蕊着生在花筒边缘。雌蕊一枚，着生在花筒底部。观察果实类型。

（4）山楂（附图 14－2） 取枝条观察。注意叶序、叶形，观察托叶形状。观察花序类型。取一朵小花放在放大镜下观察花柱特征及子房位置。横切子房，注意由几个心皮合生成，什么胎座类型。取果实横切，观察梨果特点。

根据观察解剖的植物特征，确定上述四种植物各属于蔷薇科的哪个亚科。

2. 豆科

（1）合欢 落叶乔木，二回偶数羽状复叶，头状花序排成伞房状，取 1 朵小花观察，注意其对称性及花萼、花冠的连合状况，雄蕊和心皮的数目。果实类型。树皮入药。山合欢见附图 14－3。

（2）决明 草本，观察叶的特征，花通常成对腋生，萼片 5，分离；花瓣 5，能育雄蕊 7 枚，荚果细长，近四棱形。种子称决明子。

（3）甘草（附图 14－4） 多年生草本，全株被白色柔毛和腺毛，根和根茎粗壮，外皮红褐色至暗褐色。观察其复叶类型，小叶及托叶的形态。总状花序腋生，花冠蓝紫色，雄蕊 10，二体雄蕊。荚果镰刀状或环状弯曲，密被刺状腺毛及短毛。根及根茎入药。

3. 五加科 人参（附图 14－5）：多年生草本，注意主根及根茎（芦头）的特点，掌状复叶轮生于茎端，观察复叶的形态特征。伞形花序顶生，花萼 5 齿裂；花瓣 5，淡黄绿色；雄蕊 5；花盘杯状；2 心皮合生。浆果状核果。熟时鲜红色。根及根茎、叶入药。

二、其他药用植物标本观察

1. 蔷薇科

（1）地榆（附图 14－6）　多年生草本，注意观察根的形状，奇数羽状复叶，小叶距圆状长椭圆形，边缘有锯齿，花小，密集成顶生的穗状花序，花萼紫红色，无花瓣，根作药用。

（2）贴梗海棠　落叶灌木，枝有刺，叶倒卵形，托叶大，花淡红色或白色，3～5 朵簇生，花筒钟状，梨果球形或卵形，木质，果实干燥后称为皱皮木瓜，作药用。

2. 豆科

（1）皂荚　落叶乔木，主干上部和枝条上常有圆柱形分枝棘刺，一回羽状复叶。总状花序腋生或顶生；花杂性，雄花花萼 4 裂，花瓣 4，白色或淡黄色，雄蕊 8，雌蕊退化；两性花较大，雄蕊 8，雌蕊能育。荚果扁长条状，成熟后黑棕色，被白色粉霜。棘刺及果实入药。

（2）膜荚黄芪（附图 14－7）　多年生直立草本，与甘草主要区别在于花冠淡黄色，有时近白色，荚果卵状长圆形，膜质，膨胀，被黑色短毛。根入药。

（3）苦参（附图 14－8）　落叶灌木，根圆柱状，外皮黄白色，羽状复叶，小叶狭长椭圆形至披针形；花淡黄色，蝶形；荚果线形，不开裂，先端具长喙。根入药。

3. 五加科

（1）西洋参　与人参很相似，主要区别是西洋参的总花梗上叶柄近等长，小叶片椭圆形，上面脉上几无刚毛，先端突尖。根入药。

（2）刺五加（附图 14－9）　茎枝密生细长倒刺。掌状复叶互生，小叶下面脉密生黄褐色毛。伞形花序顶生，花多而密；花萼绿色与子房合生，花各部分 5 基数。浆果状核果，紫黑色，干后有 5 棱。根及根茎入药。

（3）三七　多年生草本；主根粗，肉质，有疣状凸起的分枝。掌状复叶，小叶片上下脉上均密生刚毛。伞形花序顶生，小花多数。根入药。

【实验报告】

1. 绘一朵杏花的纵剖图，并写出花程式。
2. 举例说明蔷薇科四个亚科的主要区别点。
3. 列出豆科三个亚科的主要特征及代表植物。
4. 说明五加科的主要特征。

【思考题】

1. 把蔷薇科分为四个亚科的主要依据是什么？
2. 蝶形花冠与假蝶形花冠的主要区别有哪些？
3. 五加科与伞形科有哪些相似特征？在植物分类上两个科的亲缘关系远近如何？

Experiment 14 Rose Family (Rosaceae), Bean Family (Leguminosae) and Ginseng Family (Araliaceae)

Aim and demand

1. Grasp the main morphological characteristics of rose family (Rosaceae), bean family (Leguminosae) and ginseng family (Araliaceae).

2. Be capable to recognize the main medicinal plants of rose family (Rosaceae), bean family (Leguminosae) and ginseng family (Araliaceae).

Experiment materials

1. Rose family (Rosaceae)

1.1 Spirea subfamily (Spiraeoideae) Typical plant: the shoot, flower, fruit of Xiuxianju (*Spiraea salicifolia* L.).

1.2 Rose subfamily (Rosoideae) Typical plant: the total herbs of Longyacao (*Agrimonia pilosa* Ledeb.) with flower.

1.3 Peach subfamily (Prunoideae) Typical plant: the shoot with flower and fruit of Xing (*Prunus armeniaca* L.).

1.4 Pear subfamily (Pomoideae) Typical plant: the shoot, flower, fruit of Shanzha (*Crataegus pinnatifida* Bge.).

1.5 Other samples of medicinal plants Diyu (*Sangusorba officinalis* L.), Tiegeng-haitang (Chaenomeles speciosa).

2. Bean family (Leguminosae)

2.1 Mimosa subfamily (Mimosoideae) Typical plant: the shoot of Hehuan (*Albizia julibrissin* Durazz.) with flower and fruit.

2.2 Caesalpinia subfamily (Caesalpinoideae)

2.2.1 Typical plant the shoot of Jueming (*Cassia tora* L.) with fruit.

2.2.2 Other samples of pharmaceutical plants the shoot of Zaojia (*Gleditsia sinensis* Lam.), Wangjiangnan (*Cassia occidentalis* L.) with fruit.

2.3 Bean subfamily (Papilionoideae)

2.3.1 Typical plant the total plant of Gancao (*Glycyrrhiza uralensis* Fisch.) with flower, fruit.

2.3.2 Other samples of pharmaceutical plants Mojiahuangqi (*Astagalus membranaceus* (Fisch.) Bunge.), Kushen (*Sophora flavescens* Ait.).

3. Ginseng family (Araliaceae)

3.1 Typical plant the total herbs of Renshen (*Panax ginseng* C. A. Mey.) with flower and fruit.

3.2 Other samples of pharmaceutical plants Xiyangshen (*Panax quinquefolium*

L.), Ciwujia (*Acanthopanax senticosus* (Rupr. Et. Maxim.) Harms), Sanqi (*Panax notoginseng* (*Burk.*) F. H. Chen).

Instruments and aplliances

Magnifier, anatomical lens, dissector.

Contents and procedures

1. Observe and dissect the typical plants as follows

1.1 Rose family (Rosaceae)

1.1.1 Spirea subfamily (Spiraeoideae)

Spiraea salicifolia : observe the shoot, pay attention to the phyllotaxy, shape of leaves; and if there is stupule or not. Observe carefully to decide the type of inflorescence; Pick up a single little flower and observe with a magnifier to count the number of sepals, petals, stamens and pistils, pay attention to the shape of hypanthium; Is it perigynous flower or epigynous flower? And what type of fruit?

1.1.2 Rose subfamily (Rosoideae)

Agrimonia pilosa (Figure 14 – 1): observe the total herb with flower, dense with villies; odd – pinnately compound leaf, leaflet 5 – 7, not the same size, little alternating with large. Pay attention to the shape of leaflets and their different sizes; terminal raceme; to observe whether there is a ring of hooked hair at the top of hypanthium. Count the number of carpels and observe the type of fruit. Total herb is used in making drugs.

1.1.3 Peach subfamily (Prunoideae)

Prunus armeniaca : observe the branch, pay attention to the shape of leaves; if there is stupule or not; count the number of sepals, petals, stamens; Dissect the flower radially and pay attention to the shape of calyx tube; perianth and stamen are at the edge of hypanthium; pistil 1, at the bottom of hypanthium; observe the type of fruit.

1.1.4 Pear subfamily (Maloideae)

Crataegus pinnatifida (Figure 14 – 2): observe the shoot. Pay attention to phyllotaxy, shape of leaf; shape of stupule; type of inflorescence; observe one flower, pay attention to the position of ovary and character of style; Cut ovary transversely, to observe the number of carpels, type of placenta and characters of pome.

1.2 Bean family (Leguminosae)

Albizia julibrissin : decidulous tree; bipinnate leaf, even – pinnately compound leaf; capitulum, gather to be corymbose shape; pay attention to the symmetry of flower and how the calyx unites with corolla; count the number of stamens and carpels; to decide the type of fruit; bark is used in making drugs (Figure 14 – 3).

Cassia tora : herbs; at first to observe leaves. Flowers often axillary with another one; sepals 5, solitary; petals 5; stamens 10, 3 sterile; pod, long, near rhombus; the seed in Traditional Chinese Medicine is silktree albizzia seed (Jue – ming – zi).

Glycyrrhiza uralensis (Figure 14 – 4) : perennial herbs; the total body has white hair and glandular hairs; root and rhizome, robust, the surface is red brawn to dark brawn; observe the type of compound leaf, morphol of leaflet and stupule; axillary indefinite inflorescence; corolla blue purple; stamens 10, diadelphous stamen; pod like sickle or curve like ring, dense with thorny glandular hairs and hair; root and rhizome are used in making drugs.

1.3 Ginseng family (Araliaceae)

Panax ginseng (Figure 14 – 5): perennial herbs; study the characters of main root and rhizome; palmately compound leaf whorled; observe the morphological character of compound leaf, terminal umbel; calyx 5 – lobed; petals 5, yellowish green; stamens 5; disc likes cup; 2 carpels united; berry – like drupe, red after mature; root, rhizome and leaf are used in making drugs.

2. Observe other samples of pharmaceutical plants

2.1 Rose family (Rosaceae)

Sangusorba officinalis (Figure 14 – 6): perennial herb; pay attention to the shape of root; odd – pinnately compound leaf; spur of leaflet elliptical, at the edge is serrate; flower small, united into terminal spike; calyx purple, without petals. This root is used in making drugs.

Chaenomeles speciosa: thorny shrub; leaf obovate; large stipule; flower pink or white, clustered 3 – 5; hypanthium campanulate; spherical or ovate pome, woody; dry fruit named Zhoupimugua is used in making drugs.

2.2 Bean family (Leguminosae)

Gleditsia sinensis: decidulous tree; the up trunck and branches usually bearing stout branched spines; pinnately compound leaf; terminal or axillary indefinite inflorescence; polygamous flower, calyx of male flower 4 – lobed, petals 4, white or yellow, stamens 8, pistil reduced; large bisexual flower, stamens 8, pistils fertile. Pod is flat and long, black brown after mature, with white powder; spines and fruit are used in making drugs.

Astagalus membranaceus (Figure 14 – 7): perennial herb, erect; the main differences with *Glycyrrhiza uralensis* are that: corolla light yellow, sometimes near white; ovate pod, long orbicular, membranaceous, swelled with black hairs; root is used in making drugs.

Sophora flavescens (Figure 14 – 8): decidulous shub; terete root, the surface is yellowish white; pinnately compound leaf; leaflet from oblong to lanceolate; flower light yellow, papilionaceous corolla; linear pod, indehesicent, the apical of the pod with a long peak; root is used in making drugs.

2.3 Ginseng family (Araliaceae)

Panax quinquefolium: like *Panax ginseng* very much, the main differences are that: near equal petiole of peduncle; leaflet elliptical, the up vein without seta, the apical mucronate; root is used in making drugs.

Acanthopanax senticosus (Figure 14 – 9): shrub; ancistrus stem and branches; alternate palmately compound leaf, the lower vein of leaflet dense with yellowish brown hair; terminal umbel; flower, dense and many; green calyx unites with ovary; flower 5 – merous; berry –

like drupe, purlish black, with 5 ridges when dried. Root and rhizome are used in making drugs.

Panax notoginseng: perennial herb; thick main root, succulent, with many warty branches; palmately compound leaf; the up and the lower vein of leaflet both dense with seta; terminal umbel; little flower many; root is used in making drugs.

Experiment report

1. Draw vertical section of *Prunus armeniaca* and write it's flower formula.

2. Describe the main differences among the four subfamilies of rose family (Rosaceae) and take an example.

3. List the main characteristics of the three subfamilies of bean family (Leguminosae) and several typical pharmaceutical plants.

4. List the main characteristics of ginseng family (Araliaceae).

Questions

1. What is the base that divide rose family (Rosaceae) into four subfamilies? what is the main pharmaceutical values of rose family (Rosaceae)?

2. Describe the main differences between the papilionaceous corolla and false papilionaceous corolla.

3. List the similar characteristics between ginseng family (Araliaceae) and carrot family (Umbelliferae), how about the kinship between them?

实验十五 伞形科、木犀科、唇形科、茄科

【目的要求】

1. 掌握伞形科、木犀科、唇形科、茄科植物的主要形态特征。

2. 识别各科主要的药用植物。

【实验材料】

一、伞形科

1. 代表植物 白芷（*Angelica dahurica*（Fisch. ex Hoffm.）Benth et Hook. f.）带果全株。

2. 其他药用植物标本 柴胡（*Bupleurum chinense* DC.），蛇床（*Cnidium monnieri*（L.）Cuss.）。

二、木犀科

1. 代表植物 连翘（*Forsythia suspensa*（Thunb.）Vahl.）带花枝条及果实。

2. 其他药用植物标本 女贞（*Ligustrum lucidum* Ait.）、苦枥白蜡树（*Fraxinus rhynchophylla* Hance）带花枝条、果实。

三、唇形科

1. 代表植物 益母草（*Leonurus heterophyllus* Sweet）全株、花。

2. 其他药用植物标本 丹参（*Salvia miltiorrhiza* Bge.）、黄芩（*Scutellaria baicalensis* Georgi）、薄荷（*Metha haplocalyx* Briq.）。

四、茄科

1. 代表植物 白花曼陀罗（*Datura metal* L.）的枝、叶、花、果。

2. 其他药用植物 宁夏枸杞（*Lycium barbarum* L.）、莨菪（*Hyoscyamus niger* L.）。

【仪器用品】

放大镜、解剖镜、解剖用具。

【内容与方法】

一、观察解剖下列各代表植物

1. 伞形科 白芷：多年生草本，注意是复叶还是单叶，叶柄下部是否有膨大成囊状的叶鞘。复伞形花序，总苞片缺或 1～2，花小，花瓣白色，先端内凹。双悬果长圆形，背棱扁，侧棱翅状，棱槽中有 1 油管，合生面有 2。根入药。

2. 木犀科 连翘：落叶灌木，取一个枝条，观察茎的形状，纵剖枝条观察髓是空心还是实心，看清叶的着生方式，叶形后，取花观察。注意花为 4 基数，连合；雄蕊 2；子房上位，柱头 2 裂。蒴果木质，2 瓣裂，种子多数，说明由几个心皮构成及胎座类型。果实入药。

3. 唇形科 益母草（附图 15－1）：草本，取枝条观察茎的形状及叶序类型。注意下部、中部、上部叶的分裂有何变化。轮伞花序腋生，小苞片针刺状，花萼具 5 齿裂，裂片刺芒状。花冠唇形，粉红至淡紫色，上唇全缘，下唇 3 列；雄蕊 4，2 强；子房深 4 裂，小坚果矩圆状，褐色。全草入药。

4. 茄科 白花曼陀罗（附图 15－2）：草本，茎基部木质化，上部多分枝，单叶互生，卵形或宽卵形，基部偏斜。花单生叶腋，花萼筒状，先端 5 裂，与果实对照看，花后于基部周裂而脱落，基部宿存，花冠漏斗状，白色，雄蕊 5，子房不完全 4 室，观察果实属何种类型，花称洋金花入药。

二、其他药用植物标本观察

1. 伞形科

（1）柴胡（附图 15－3）　主根坚硬，纤维性强。注意茎上部分枝成“之”字形弯曲，叶线状披针形或倒披针形，全缘。平行脉序。复伞形花序，总苞片 2～3，狭披针形，花黄色，双悬果长圆形。根作药用。

（2）蛇床　一年生草本，茎多分枝。注意观察叶片的分离情况及裂片的形状。复伞形花序，具总苞片和小苞片，花白色。双悬果长圆形，分果背腹压扁。果实入药。

2. 木犀科

（1）女贞　常绿乔木，单叶对生，革质，圆锥花序顶生，取花观察，注意花的各部分基数，花冠漏斗状，4 裂，白色，雄蕊 2。横切雌蕊，可见由两心皮构成，2 室，核果。果实入药。

（2）苦枥白蜡树　落叶乔木，单数羽状复叶，对生，小叶 5～7，卵形，具尾尖，圆锥花序顶生，花 4 基数，雄花密集，花萼钟状，无花瓣，雌蕊疏离，注意观察雌花的柱头几裂，及果实的形态和类型。树皮入药。

3. 唇形科

（1）丹参（附图 15－4）　多年生草本，全株密被淡黄色的柔毛及腺毛，根粗大，表面砖红色。看清茎的形态，叶的形态及着生方式后，分辨是什么叶序类型，注意花冠的类型，上唇近盔状，下唇 3 裂，雄蕊 2，着生于下唇的中部，药隔伸长呈杠杆状，上端药室发育，下端药室退化，子房 4 深裂，花柱着生子房基部，小浆果长圆形。根入药。

（2）黄芩（附图 15－5）　多年生草本，主根粗壮，折断后观察断面颜色；茎丛生，基部伏地；注意观叶的形状及着生的方式；总状花序顶生，花偏向一侧，花萼，花冠均为二唇型，注意花萼的上唇背部具盾状物。根入药。

（3）薄荷（附图 15－6）　多年生草本，全株有清凉的香气，观察叶形，叶序，注意花序的类型。全草入药。

4. 茄科

（1）宁夏枸杞　灌木或呈小乔木状，枝条具棘刺，叶生卵形至卵状披针形，花冠筒长于花冠裂片，裂片无缘毛。果实，根皮入药。

（2）莨菪　二年生草本，全株被黏性腺毛和柔毛，具特殊气味，根粗壮，基生叶丛生，茎生叶互生，叶长圆形，花萼筒状钟形，5 浅裂，花冠漏斗形，黄色，蒴果顶端盖裂，藏于宿萼内。种子圆肾形有网纹。种子入药。

【实验报告】

1. 绘白芷花的纵剖图及分果的横切面简图，并注明各部分名称。
2. 绘连翘（金钟花）花的解剖图并注明个部分的名称。
3. 写出唇形科的主要特征，并列出常用的药用植物及药用部位。
4. 绘出观察到的白花曼陀罗花的纵剖图并注明个部分的名称。
5. 写出茄科的主要特征，并列出常用的药用植物及药用部位。

【思考题】

1. 双悬果主要的构造有哪些？
2. 比较唇形科和马鞭草科的异同点。
3. 举例说明茄科植物是与人类关系比较密切的一类植物。

Experiment 15 Carrot Family (Umbelliferae), Olive Family (Oleaceae), Mint Family (Labiatae), Nightshade Family (Solanaceae)

Aim and demand

1. Grasp the main morphological characters of carrot family (Umbelliferae), olive family (Oleaceae), mint family (Labiatae), nightshade family (Solanaceae)

2. Be capable to recognize the main medicinal plants of carrot family (Umbelliferae), olive family (Oleaceae), mint family (Labiatae), nightshade family (Solanaceae)

Experiment materials

1. Carrot family (Umbelliferae)

1.1 Typical plant the total herbs of Baizhi (*Angelica dahurica* (Fisch. ex Hoffm.) Benth et Hook. f.) with fruit.

1.2 Other samples of pharmaceutical plants Chaihu (*Bupleurum chinense* DC.), Shechuang (*Cnidium monnieri* (L.) Cuss).

2. Olive family (Oleaceae)

2.1 Typical plant the shoot with flowers and fruit of Lianqiao (*Forsythia suspensa* (Thunb.) Vahl.).

2.2 Other samples of pharmaceutical plants the shoot with flowers and fruit of Nvzhen (*Ligustrum lucidum* Ait.) and Kulibailashu (*Fraxinus rhynchophylla* Hance).

3. Mint family (Labiatae)

3.1 Typical plant the total herbs, flower of Yimucao (*Leonurus heterophyllus* Sweet).

3.2 Other samples of pharmaceutical plants Dansen (*Salvia miltiorrhiza* Bge.), Huangqin (*Scutellaria baicalensis* Georgi), Bohe (*Metha haplocalyx* Briq).

4. Nightshade family (Solanaceae)

4.1 Typical plant the shoot, leaf, flower and fruit of Baihuamantuoluo (*Datura metal* L.).

4.2 Other samples of pharmaceutical plants the shoot, and fruit of Ningxiagouqi (*Lycium barbarum* L.), the shoot, leaf, flower and fruit of Langdang (*Hyoscyamus niger* L.).

Instruments and appliances

Magnifier, anatomical lens, dissector.

Contents and procedures

1. Observe and dissect the typical plants as follow

1.1 Carrot family (Umbelliferae)

Angelica dahurica: perennial herbs; observe the leaves to decide whether it is simple leaf or compound leaf? if there is leaf sheath that swell like a sac at the lower part of petiole? compound umbel, non - involucre or involucre 1 - 2; small flower, petals white, the apical half invaginates; cremocarp long orbicular; dorsal edge flat, lateral edge likes wing; 1 oil tube in the groove of edge, and 2 in association; root is used in making drugs.

1.2 Olive family (Oleaceae)

Forsythia suspense: a decidulous shrub. Observe the shape of stem and whether the pith is full or hollow; describe the phyllotaxy and the type of leaf; flower is 4 – merous, united, 2 stamens, superior ovary, and 2 – lobed stigma, capsule is woody, 2 – valved, many seeds; count the number of carpels and study the type of placenta. Fruit is used in making drugs.

1.3 Mint family (Labiatae)

Leonurus heterophyllus (Figure 15 – 1): herb; observe the type of phyllotaxy and shape of stem, pay attention to the difference of leaf division from the up to the middle and the low of stem. verticillaster is axillary, acupuncture bractlet, 5 – short teeth calyx, awned lobe, labiate corolla with the colour from pink to purple pink, entire upper lip, the 3 – lobed lower lip, 4 stamens, didynamous, deeply 4 – lobed ovary. nutlet is orbicular, brown. The herb is used in making drugs.

1.4 Nightshade family (Solanaceae)

Datura metel herb (Figure 15 – 2): the base of stem is lignified and the upper is multi-ramose; simple leaf is alternative, ovate or broad ovate, obique base; flower is united from leaf axil, tubular calyx, lysising 5 parts at the tip, funnel – shaped corolla, white, 5 stamens, incomplete 4 – roomed ovary. flower fall out after basal circumscissile and base is persistent. Observe the type of fruit. Flower called Yang – Jinhua is used in making drugs.

2. Oberve other samples of pharmaceutical plants

2.1 Carrot family (Umbelliferae)

2.1.1 *Bupleurum chinense* (Figure 15 – 3) sturdy main root, fiberous; pay attention to the up of the stem zigzag likes; linear leaf lanceolate or oblanceolate, entire; paralle venation; compound umbel; involucre 2 – 3, thin lanceolate; yellow flower; cremocarp, long orbicular; root are used in making drugs.

2.1.2 *Cnidium monnieri* annual herb; stem with many branches; observe the division of leaf and the shape of lobe; compound umbel, with invooucre and braclet; white flower; cremocarp, long orbicular, each schizocarp with flat back and the ventral; fruit is used in making drugs.

2.2 Olive family (Oleaceae)

2.2.1 *Ligustrum lucidum* an aiphyllium. Simple Leaves are opposite, coriaceous; panicle is terminal, 4 – lobes, white, and funnel – shaped corolla, and 2 stamcns, pay attention to the basal number is 4; take the transverse ovary and observe that ovary is composed of two carpels with 2 rooms; drupe fruit. Fruit is used in making drugs.

2.2.2 *Fraxinus rhynchophylla* a defoliate arbor. Odd – pinnately compound leaf is opposite, ovaliform, 5 – 7 foliole with caudal tip; panicle is terminal, flower 4 – merous, congested male flower, campanulate calyx, non – petal, and remote stamens. Observe the number of parts of stigma of female flower and the shape and type of fruit. Bark is used in making drugs.

2.3 Mint family (Labiatae)

2.3.1 *Salvia miltiorrhiza* (Figure 15 – 4) an etesiae, densely covered by flavescent vellus hair and glandular hairs; root is grossus, bolarious surface. Discriminate the type of phyllotaxy after observing the shape of stem and the shape and inserted form of leaf, pay attention to the type of corolla, nearly helmet – shaped upper lip, 3 – lobed lower lip, 2 stamens, inserting in the middle of the lower lip, elongated and clavillosus connective, auxetic anther chamber lying superior extremity and retrogressive one lying inferior extremity, deeply 4 – lobed ovary, gynobasic stylar. Small berry is prolate spheroidal. Root is used in making drugs.

2.3.2 *Scutellaria baicalensis* (Figure 15 – 5) an etesiae, with stout main root. Ascertain the colour of bisect. Stem is rosette, paving base. Pay attention to the shape and inserted form of leaves; racemose inflorescence is terminal, flower incline to one side, bilabiate calyx and corolla. Pay attention to the dorsum of upper lip of calyx with scutellate thing. Root is in drug use.

2.3.3 *Metha haplocalyx* (Figure 15 – 6) an etesiae, emitting cool and refreshing scent; observe phylliform and phyllotaxy; pay attention to the type of infloresceuce. The total herb is used in making drugs.

2.4 Nightshade family (Solanaceae)

2.4.1 *Lycium barbarum* a shrub or small tree, with thorny shoot. Leaves are oval to ovalifrom – lanceolate; Corolla tube is longer than collroa lobe with non – marginal seta. Fruit and the cortex of root are used in making drugs.

2.4.2 *Hyoscyamus niger* a biennial grass. Covered viscose rellus hair and glandular hair, emitting a special odor, with stout root; Basal leaves are rosette, but stem leaves are alternative, leaves are prolate spherical; Calyx is tubular, campanulate dividing 5 – short parts; capsular fruit is dehiscent by lid at the tip, covered in the persistent calyx. Seed is used in making drugs.

Experiment report

1. Draw a vertical section of flower of *Angelica dahuriea* and the transverse section of schizocarp.
2. Draw the flower of *Forsythia suspense* and note the name of every section.
3. List the main characteristics of the mint family (Labiatae) and write the commonly used pharmaceutical plants and pharmaceutical section.
4. Draw the vertical section of flower of *Datura metel L.* and note the name of every parts.
5. List the main characters of nightshade family (Solanaceae) and write the commonly used pharmaceutical plants and pharmaceutical section.

Questions

1. What is the main structure of cremocarp?
2. Compare the similarities and difference between Labiatae and Verbenaceae.

3. Illustrate and give examples that the plant of nightshade family (Solanaceae) has a fairly close relationship with human being.

实验十六　玄参科、桔梗科、菊科

【目的要求】

1. 掌握玄参科、桔梗科、菊科植物的主要形态特征。
2. 识别各科主要的药用植物。

【实验材料】

一、玄参科

1. 代表植物　地黄（*Rehmannia glutinosa*（Gaertn.）Libosch.）带花全株、果。

2. 其他药用植物标本　玄参（*Scrophularia ningpoensis* Hemsl.）带花全株、果；洋地黄（*Digitalis purpurea* L.）带花全草、果实；北玄参（*Scrophularia buergeriana* Miq.）带花全草、果实。

二、桔梗科

1. 代表植物　桔梗（*Platycodon grandiflorum*（Jacq.）A. DC.）带花全株、果。

2. 其他药用植物标本　沙参（*Adenophora stricta* Miq.）带花植株、根；党参（*Codonop-sis pilosula*（Franch.）Nannf.）带花植株、根；半边莲（*Lobelia chinensis* Lour.）全株。

三、菊科

1. 管状花亚科

（1）代表植物　红花（*Carthamus tinctorius* L.）带花全草。

（2）其他药用植物标本　菊花（*Dendranthema morifolium*（Ramat.）Tzvel.）带花全草；北苍术（*Atractylodes chinensis*（DC.）Koidz.）带花全草；白术（*Atractylodes macrocephala* Koidz.）植株；紫菀（*Aster tataricus* L. f.）带花植株。

2. 舌状花亚科

（1）代表植物　蒲公英（*Taraxacum mongolicum* Hand. - Mazz.）带花全草、果。

（2）其他药用植物　苦苣菜（*Sonchus oleraceus* L.）全草。

【仪器用品】

放大镜、解剖镜、解剖用具。

【内容与方法】

一、观察解剖下列各代表植物

1. 玄参科　地黄（附图 16－1）：取地黄带花全株观察。全体密被灰白色柔毛及腺毛，观察是根状茎、块茎还是块根？在看清叶的着生方式、叶形、花序后取花观察，注意花冠的形状、颜色、上下唇裂片的数目、上唇是否反卷、雄蕊是否为二强。横切子房，观察子房室、胚珠的数目及胎座类型。观察果实类型、颜色，种子数目、形状和颜色。块根入药。

2. 桔梗科 桔梗（附图16－2）：取桔梗植株观察叶序、叶形及根的特征。鲜标本折断茎叶，看是否有乳汁流出？取花观察，花5基数，注意观察花冠的形状、子房半下位。横切子房，观察子房室数、胎座类型。蒴果顶端5瓣裂。根入药。

3. 菊科

（1）管状花亚科　红花（附图16－3）：取红花植株观察根、茎、叶的外部形态，注意叶序、叶形，是否抱茎。取花序观察总苞片的形状、列数、花托的形状。观察小花的类型，花萼的有无，是否全为管状花。解剖小的管状花，注意其是两性花、单性花还是中性花。再观察心皮数目、子房类型及雄蕊，是否为聚药雄蕊。果实为瘦果。折断鲜植物看是否有乳汁流出。管状花入药。

金盏菊：观察方法同红花，其缘花为舌状花，盘花为管状花。

（2）舌状花亚科　蒲公英（附图16－4）：取蒲公英植株，按照红花的观察方法操作观察，注意其与红花的区别点。花全部为舌状花，植物体有白色乳汁，瘦果顶端有冠毛。全草入药。

二、其他药用植物标本观察

1. 玄参科

（1）玄参　块根数条，茎方形，下部叶对生，上部叶有时互生，叶片卵形至披针形，聚伞花序合成大而疏散的圆锥花序，花萼5裂，花冠褐紫色，雄蕊4枚、2强，蒴果卵形。根入药。

（2）北玄参　与玄参区别，聚伞花序缩成穗状，花冠黄绿色。

（3）紫花洋地黄　与地黄区别，基生叶多数呈莲座状，花萼钟状，5裂深至基部，蒴果卵形，顶端尖，密被腺毛。

2. 桔梗科

（1）沙参（附图16－5）　根肥大，圆锥形，叶互生。折断鲜标本观察有无乳汁？观察花序类型，取花观察，花5基数，钟状花冠，注意花柱基部有杯状花盘，子房下位。横切子房，观察其由几枚心皮组成，胎座类型。果实为蒴果。

（2）党参（附图16－6）　草质藤本，根肉质，叶互生，卵状心形。花冠钟状，黄绿色，具淡紫色斑点。蒴果顶端瓣裂。根入药。

（3）半边莲　小草本，具乳汁，叶互生，花冠粉红色近唇形，裂片偏向一侧，上唇2裂，下唇3裂，全草药用。

3. 菊科

（1）菊花　全株被白色柔毛，叶卵形至披针形，互生，边缘有粗大锯齿或深裂成羽状。取头状花序仔细观察，总苞片多层，缘花舌状，雌性，盘花管状，两性，黄色，具托叶，瘦果无毛。花序入药。

（2）北苍术（附图16－7）　根状茎结节状，断面具有红色油点，叶互生、革质，下部叶常3裂，上部常不裂，叶缘具刺状齿。花冠白色，头状花序全为管状花。根茎入药。

（3）白术　与北苍术的区别，根状茎略成拳状，叶片3裂有长柄，花冠紫色。根茎入药。

（4）紫菀（附图16-8）　根状茎短，簇生多数细根。基生叶匙形，茎生叶披针形。头状花序排列成复伞房状。缘花舌状，蓝紫色，盘花管状，黄色。根状茎及根入药。

（5）苦苣菜　根纺锤形，茎上部有的具腺毛，叶羽状深裂或大头羽状半裂，头状花序，全为舌状花。全草入药。

【实验报告】

1. 列出玄参科的主要特征，并列举数种常见药用植物及药用部位。
2. 绘观察解剖的桔梗花的纵剖图，并写出花程式。
3. 绘菊科药用植物的纵剖面的一朵管状两性花、一朵舌状雌花、一朵舌状两性花的纵剖面图并注明各部分名称。
4. 说明菊科头状花序的结构和小花有哪些类型。

【思考题】

1. 菊科的主要特征有哪些？分成两个亚科有哪些依据？
2. 菊科植物有哪些特征可以说明其进步性，这些特征在适应环境时有什么意义？
3. 比较玄参科、唇形科的主要异同点。

Experiment 16　Figwort Family (Scrophulariaceae), Bellflower Family (Campanulaceae), Compositae Family (Compositae)

Aim and demand

1. Grasp the main morphological characteristics of the figwort family (Scrophulariaceae), bellflower family (Campanulaceae), compositae family (Compositae).

2. Be able to identify the main pharmaceutical plants of figwort family (Scrophulariaceae), bellflower family (Campanulaceae), compositae family (Compositae).

Experiment materials

1. Figwort family (Scrophulariaceae)

1.1 Typical plant　the total herbs of Dihuang (*Rehmannia glutinosa* (Gaertn.) Libosch.) with flowers and fruits.

1.2 Other samples of pharmaceutical plants　Xuanshen (*Scrophularia ningpoensis* Hemsl) with flowers and fruits; Yangdihuang (*Digitalis purpurea* L.) and Beixuanshen (*Scrophularia buergeriana* Miq.).

2. Bellflower family (Campanulaceae)

2.1 Typical plant The total herbs of Jiegeng (*Platycodon grandiflorum* (Jacq.) A. DC.) with flowers and fruits.

2.2 Other samples of pharmaceutical plants Total herbs of Shashen (*Adenophora stricta* Miq.), Dangshen (*Codonopsis pilosula* (Franch.) Nannf.), Banbianlian (*Lobelia chinensis* Lour.).

3. Compositae family (Compositae)

3.1 Tubular – flowered subfamily (Tubuliflorae)

3.1.1 Typical plant total herbs of Honghua (*Carthamus tinctorius* L.) with flowers.

3.1.2 Other samples of pharmaceutical plants total herbs of Juhua (*Dendranthema morifolium* (Ramat.) Tzvel.), BeiChangzhu (*Atractylodes* chinensis (*DC.*) *Koidz*), *Ziwan* (Aster tataricus *L. f.*) *with flowers*, *total herbs of Baizhu* (Atractylodes macrocephala *Koidz.*).

3.2 Ligulate – flowered subfamily (Liguliflorae)

3.2.1 Typical plant total herbs of Pugongying (*Taraxacum Mongolicum* Hand – Mazz.).

3.2.2 Other samples of pharmaceutical plants total herbs of Kujucai (*Sonchus oleraceus* L.).

Instruments and aplliances

Magnifier, anatomical lens, dissector.

Contents and procedures

1. Observe and dissect the typical plants as follow

1.1 Figwort family (Scrophulariaceae)

Rehmannia glutinosa (Figure 16 – 1): total herb with a lot of grayish white villi and glandular hairs; to decide whether it is rhizome, tuber or root tuber? After having a good look and making sure the phyllotaxy, shape of leaf and the type of inflorescence of flower, to study the shape and color of corolla, number of upper and lower lips, if the upper lip is trans – curl or not, if stamen is didynamous stamen or not? Transverse cut the ovary to count the number of ovary cells, ovules and decide which type of placenta it is. observe the shape and color of fruit, so do the number, shape and color of seed. Root tuber is used in making drugs.

1.2 Bellflower family (Campanulaceae)

Platycodon grandiflorum (Figure 16 – 2): observe the total herb, get the characters of phyllotaxy, shape of leaf and root; snap the stem or leaf in two parts, and see if there is milky juice or not; flower 5 – merous; pay attention to the shape of corolla; half – inferior ovary, transvers cut the ovary to understand the numbers of ovary cells and type of placenta; capsule 5 – celled, splitting at the top by means of valves. Root is used in making drugs.

1.3 Compositae family (Compositae)

1.3.1 Tubular – flowered subfamily (Tubuliflorae)

Carthamus tinctorius (Figure 16 – 3): observe the total herb to grasp the morphological characters of root, stem, leaf and flower; pay attention to phyllotaxy, shape of leaf, receptacle and involucre, amplexicaul leaf or not? observe the type of little flower to decide if there is calyx, if they are all tubular - flowers or not; dissect one of little tubular - flowers and observe whether it is bisexual flower, unisexual flower or asexual flower; then to count the number of carpels, decide the type of ovary, and acertain if stamen is synantherous stamen; fruit, achene nulet; snap the stem in two parts, and see if there is milky juice; tubular - flower is used in making drugs.

1.3.2 Ligulate - flowered series (Liguliflorae)

Taraxacum Mongolicum (Figure 16 - 4): do as above about *Carthamus tinctorius*, but should distinguish it from *Carthamus tinctorius*, all of the flowers are ligulate flowers; with milky juice; there are pappus at the top of nutlet; This total herb is used in making drugs.

2. Oberve other samples of pharmaceutical plants

2.1 Figwort family (Scrophulariaceae)

2.1.1 *Scrophularia ningpoensis* root tuber; square stem; lower leaf opposite, upper leaf somewhat alternate; leaf, ovate or lanceolate; definite inflorescence unites into panicle; calyx 5 - lobed; corolla purplish brown; 4 - stamen, didynamous; capsule ovate, at the top with glandular hairs. Root is used in making drugs.

2.1.2 *Scrophularia buergeriana* distinguish it from *Scrophularia ningpoensis* that: corolla, yellow green; definite inflorescence unites into spike.

2.1.3 *Digitalis purpurea* distinguish it from *Rehmannia glutinosa* that: basal leaf usually rosella; campaniform calyx 5 - lobed partite into base; capsule ovate; the top erect with glandular hairs.

2.2 Bellflower family (Campanulaceae)

2.2.1 *Adenophora stricta* (Figure 16 - 5) root hypertrophic, like cone; leaf Alternate; snap the fresh plant to see if there is milky juice or not; observe type of inflorescence; flower 5 - cardinal; campaniform corolla; floral disc is like cup; inferior ovary; from transverse section we may know number of carpel and type of placenta; fruit, capsule.

2.2.2 *Codonopsis pilosula* (Figure 16 - 6) herb with twining stem; root hypertrophic; leaf alternate, ovate like cordate; campaniform corolla, yellow green, with purle pink puncta; capsule valved at the top; root is used in making drugs.

2.2.3 *Lobelia chinensis* herb, with milky juice; leaf alternate; corolla usually similar to 2 - lipped, pink red, upper 2 - lobed, lower 3 - lobed; calyx split down into one side. Total herbs is used in making drugs.

2.3 Compositae family (Compositae)

2.3.1 *Dendranthema morifolium* the total herbs has white villi; leaf alternate, ovate or lanceolate, and the margin is serrate or pinnatipartite; heads, several layers of involucres; ray flower, ligulate - flower, female; discoid flower, tubular - flower, bisexual, yellow; with stipules; nutlet with no hair; inflorescence is used in making drugs.

2.3.2 *Atractylodes chinensis* (Figure 16 – 7) rhizome tuberal shape; in transverse section has red brown oil spot; leaf, alternate, herbaceous; upper leaf 3 – lobed; lower leaf non lobed; leaf margin has eyelash – like tooth; corolla, white; capitulum with all tubular – flowers; rhizome is used in making drugs.

2.3.3 *Atractylodes macrocephala* the differences between *Atractylodes macrocephala* and *Atractylodes chinensis* are that : rhizome of *Atractylodes macrocephala* is like fist; leaf, 3 – lobed, with long petiole; corolla, purple; rhizome is used in making drugs.

2.3.4 *Aster tataricus* (Figure 16 – 8) short rhizome, fascicled fibrous root; basal leaf, spoon – shaped, normal leaf lanceolate; capitulum range into compound corymb; ray flower, ligulate – flower, blueish purple; discoid flower is tubular – flower, yellow; rhizome and root are used in making drugs.

2.3.5 *Sonchus oleraceus* root like spindle; upper of stem has glandular hairs; leaf pinnatipartite or pinnatifid; capitulum with all ligulate – flowers; total herbs is used in making drugs.

Experiment report

1. List the main characteristics of the figwort family (Scrophulariaceae) and several common pharmaceutical plants and pharmaceutical section.

2. Draw the vertical section of flower of *Platycodon grandiflorum* and write it's flower formula.

3. Draw the vertical sections of flowers of pharmaceutical plants of compositae family as follow: a tubular bisexual flower, a ligulate female flower, a ligulate bisexual flower and note the name of every section.

4. Describe the structure of capitulum and type of little flower.

Questions

1. List the main characteristics of the compositae family (Compositae), what are the bases that divide compositae family (Compositae) into tubular – flowered series (Tubuliflorae) and ligulate – flowered series (Liguliflorae)?

2. Which characteristics of compositae family (Compositae) plants can verify it has advances , and what good are they to adopt to the environment ?

3. Compare the figwort family (Scrophulariaceae) with nightshade family (Solanaceae), write the similarities and the differences between them.

实验十七　禾本科、天南星科、百合科

【目的要求】

1. 掌握禾本科、天南星科、百合科植物的主要形态特征。

2. 识别各科主要的药用植物。

【实验材料】

一、禾本科

1. 代表植物　薏苡（*Coix lacryma - jobi* L. var. *ma - yuen*（Roman.）Stapf.）的带花全草及果实。

2. 其他药用植物标本　芦苇（*Phragmites communis* Trin.）的带花全草。

二、天南星科

1. 代表植物　半夏（*Pinellia ternate*（Thunb.）Breit.）带花序植株、果序。

2. 其他药用植物标本　天南星（*Arisaema consanguineum* Schott）带花植株、果序；石菖蒲（*Acorus tatarinowii* Schott.）带花植株。

三、百合科

1. 代表植物　百合（*Lilium brownii* F. E. Brown. var. *Viridulum* backer.）带花植株及鳞茎。

2. 其他植物标本　黄精（*Ploygonatum sibiricum* Redoute.）全草；麦冬（*Ophiopogon japonicus*（L. f.）Ker - Gawl.）植物标本；暗紫贝母（*Fritillaria unibracteata* Hsiao et K. C. Hsio）带花植株。

【仪器用品】

放大镜、解剖镜、解剖工具。

【内容与方法】

一、观察解剖下列各代表植物

1. 禾本科　薏苡（附图 17 - 1）：草本，杆直立，基部节上生根，叶互生，叶鞘与叶片间具白色膜状的叶舌，叶片长披针形，基部鞘状抱茎，总状花序腋生，小穗单性；雄小穗排列于花序上部，雌小穗生于花序下部，包藏于骨质总苞中，总苞在果实成熟时坚硬而光滑、质脆易破碎，内含颖果，种子称薏苡仁入药。

2. 天南星科　半夏（附图 17 - 2）：草本，全株光滑无毛，块茎扁球形，成年植株叶 3 全裂，叶柄近基部有珠芽，观察佛焰苞的形态和颜色，剥开佛焰苞观察肉穗花序上的雌蕊和雄蕊的形态，雄花在上部，每雄花仅有两个雄蕊，雌花在下部，子房 1 室，胚珠 1 枚，浆果，块茎入药。

3. 百合科　百合（附图 17 - 3）：草本，具鳞茎，叶互生，花乳白色大型，喇叭状，注意观察花被片基部具蜜槽。花被片先端外弯，花冠喉部淡黄色，雄蕊六枚，花药丁字形。花柱细长，柱头 3 裂。取子房，横切，观察子房室数及胎座类型，蒴果。鳞茎入药。

二、其他药用植物标本的观察

1. 禾本科　芦苇：草本，根茎粗壮，叶片广披针形至宽条形，叶舌有毛，圆锥状花序顶生，微垂头，分枝纤细，呈扫帚状，棕紫色，根茎入药。

2. 天南星科

（1）天南星（附图 17－4）　草本，块茎扁球形，叶 1～3 枚，叶片辐射状全裂，具 10～24 裂片，裂片披针形。观察佛焰苞的形态和颜色，管部圆柱形，喉部戟形不闭合；肉穗花序附属体棒状；花单性异株，总花梗短于叶柄；雄花具雄蕊 4～6，花丝愈合，雌花密集，每花具 1 雌蕊。浆果红色，聚合成穗状。块茎入药。

（2）石菖蒲　草本，全株具浓烈的芳香气，根状茎平卧，叶基生，无柄，叶片狭条形，无中肋，佛焰苞与叶同形同色，不包被花序，花两性。根状茎入药。

3. 百合科

（1）黄精　草本，根状茎近圆锥状，叶通常 4～6 片轮生，叶片条状披针形，先端卷曲。花腋生，取花观察，花被片合成管状，6 裂，用放大镜观察可见裂片先端具乳突。雄蕊 6，子房上位，横切子房，观察子房室数，胎座类型，浆果。根茎入药。

（2）麦冬（附图 17－5）　草本，具多数横生的地下块茎，叶基生，细条形。总状花序比叶短，花稍下垂，雄蕊 6，花柱粗短，基部宽阔，略呈圆锥形，子房半下位，横切子房，观察子房室数和胎座类型。块根入药。

（3）暗紫贝母　草本，鳞茎球形或圆锥形，外面的鳞叶两枚，大小悬殊。茎生叶条形或条状披针形，先端不卷曲，苞叶（1～2）形同叶片，花深紫色，内面通常黄绿色，蒴果。鳞茎入药，是川贝母中“松贝”的主要来源之一。

【实验报告】

1. 写出常用的禾本科药用植物名称及药用部位。
2. 列出天南星科植物的主要特征，并列表比较天南星和半夏在形态上有何不同。
3. 绘百合花的解剖图，注明各部分的名称，并写出花程式。
4. 写出百合科植物的主要特征。

【思考题】

1. 禾本科植物的主要特征有哪些？禾亚科和竹亚科有哪些区别？
2. 中药贝母的原植物有哪些种类？在鳞茎的形态上如何区别它们？

Experiment 17　Grass family（Gramineae），Aroid family（Araceae），Lily family （Liliaceae）

Aim and demand

1. Grasp the main morphological characteristics of the grass family（Gramineae），aroid family（Araceae）and lily family（Liliaceae）.

2. Recognize the main pharmaceutical plants of grass family（Gramineae），aroid family（Araceae）and lily family（Liliaceae）.

Experiment materials

1. Grass family (Gramineae)

1.1 Typical plant the whole herb with flowers and fruits of Yiyi (*Coix lacryma-jobi* L. var. *ma-yuen* (Roman.) Stapf.).

1.2 Other samples of pharmaceutical plants the whole plant with flowers of Luwei (*Phragmites communis* Trin).

2. Aroid family (Araceae)

2.1Typical plant infructescence of the whole plant with inflorescence of Banxia (*Pinellia ternata* (Thunb.) Breit).

2.2 Other samples of pharmaceutical plants the whole plant with flowers and infructescence of Tiannanxing (*Arisaema consanguineum* Schott). The whole plant with flowers of Shichangpu (*Acorus tatarinowii* Schott).

3. Llily family (Liliaceae)

3.1 Typical plant The whole plant with flowers and bulb of Baihe (*Lilium brownii* F. E. Brownvar. var. *Viridulum backer*).

3.2 Other samples of pharmaceutical plants The whole plant of Huangjing (*Ploygonatum sibiricum* Delar ex Red), the specimen of plant of Maidong (*Dophiopogon japonicus* (L. F.) ker-Gawl), the whole plant with flowers of Anzibeimu (*Fritillaria unibracteata* Hsiao et K. C. Hsio).

Instruments and appliances

Magnifier, anatomical lens, dissector.

Contents and procedures

1. Observe and dissect the typical plants as follows

1.1 Grass family (Gramineae)

Coix lacryma-jobi var. *ma-yuen* (Figure 17-1): a herb, with erect stem and roots growing on basal node. Leaves are alternate, point at the apex, sheaded base more or less indented and clasping the stem. Indefinite inflorescence is axillary, unisexual spikelet. Male spike lets are on the top of inflorescence, female spike lets are in the lower part of the inflorescence and enclosed in a bony, glossy, head like involucels. As ripe fruits, involucres become hard, smooth, easily obtrite, containing one grain. Seed called Yiyiren is used in making drugs.

1.2 Aroid family (Araceae)

Pinellia ternata (Figure 17-2): a herb, smooth, non-hair, with globose depressed tuber. As growed, leaves are divided into 3 lobes, nearly basal pedicel with bubil. Observe the shape and color of spathe and the shape of pistil and genitalia on the spadix through peeling the spathe, upper male flowers, every male flower with only 2 genitalia, lower female flowers 1-roomed ovary and 1 ovule. Berry. Tuber is used in making drugs.

1.3 Lily family (Liliaceae)

Lilium brownii var. *Viridulum* (Figure 17-3): a herb, with bulb. Leaves are alternate.

Flowers are large, galochrous, and trumpet - shaped, tapals which top is outer - curved, corolla which throat is ochroleucous, 6 genitalia, versatile anther, slender stylar, stigma dividing 3 parts. Pay attention to basal tapal with honey tank. Take ovary to transverse and to observe the number of the locule and type of placenta. Capsular fruit. Bulb is used in making drugs .

2. Observe other samples of pharmaceutical plants

2. 1 Grass family (Gramineae)

Phragmites communis: a grass, with a thick, creeping root - stock. Leaves are wide - lanceolate, broad - shoot, trichomatic ligules. Flowers are terminal, arranged in little nutant, slend branched, purplish brown, feathery, bump like panicles. Rhizome is used in making drugs .

2. 2 Aroid family (Araceae)

2. 2. 1 *Arisaema consanguineum* (Figure 17 - 4) a herb, with globosely depressed tuber. Leaves are usually 1 - 3, radial, divided into 10 - 24, lanceolate lobes. Observe the shape and colour of spathe, cylindrical tube, hastate non - inapertus throat, spadix is stick, unisexual flower diecious, total pedicle is shorter than petiole, staminate flowers, with stamens 4 - 6, and healing filament; pistillate flower is congested , and every small flower with 1 pistil. Red berry. Tuber is used in making drugs.

2. 2. 2 *Acorus tatarinowii* a herb, emitting vigorous aromatic odor, with prostrate rhizome. Basal leaves are no pedicel and rib, steno - shoot - shaped blades. The shape and colour of spathe is the same as leaf, spathe not wrap the non - custodies inflorescence. Flowers are bisexual. Rhizome is used in making drugs.

2. 3 Lily family (Liliaceae)

2. 3. 1 *Polygonatum sibiricum* a herb, with nearly conical rhizome. Leaves are usually cyclic in 4 - 6 lobes, linear - lanceolate, curved top. Flowers are axillary, tapal growing tubes together and lysising 6 parts, lobe with mamillary process observed through megaloscope, 6 genitalia, and superior ovary. Take transverse ovary and observe the number of locule and type of placenta. Berry. Rhizome is used in making drugs.

2. 3. 2 O*phiopogon japonicus* (Figure 17 - 5) a herb, with numerous hemitropous, hypogenous tubers. Leaves are basal, slender shoot - shaped. Racemose inflorescence is shorter than leaf, a little cernuous flowers, 6 genitalia, stumpy - formed stellar, board base more or less like cone shape, half - inferior ovary. Take transverse ovary and observe the number of locule and type of placenta. Root and tuber are used in making drugs.

2. 3. 3 *Fritillaria unibracteata* a herb, with spherical or conical bulb, 2 outer scale leaves in mainly different size. Leaves are shoot - shaped, linear - lanceolate, top is non - curved. The shape of 1 - 2 bracteal leaves are the same as leaves. Flowers are profoundly violet, medial surface is usually flavo - green. Capsular fruit. Bulb is in drug use, which is one of main sources of "Songbei" among the chuan fritillaria (Chuanbeimu) .

Experiment report

1. Write commonly used pharmaceutical plants and pharmaceutical section of grass family

(Gramineae).

2. List the main characteristics of the aroid family (Araceae), compare the morphological characteristics between *Pinellia ternate* and *Arisaema consanguineum*.

3. Draw a flower of *Lilium brownii* Var. *viridulum*, note the name of every section and write flower formula.

4. List the main characteristics of the lily family (Liliaceae).

Questions

1. What is the main characteristics of grass family (Gramineae), and what is the difference between Agrostidoideae and Bambusoideae?

2. How many kinds of original plants of Traditional Chinese Medicine - Fritillaria and there and how can one distinguish them through the shape of bulb?

实验十八 薯蓣科、鸢尾科、姜科、兰科

【目的要求】

1. 掌握薯蓣科、鸢尾科、姜科、兰科植物的主要特征。
2. 识别各科主要的药用植物。

【实验材料】

一、薯蓣科

1. 代表植物 薯蓣（*Dioscorea opposita* Thumb.）的带花藤条一段及根茎。

2. 其他药用植物 穿龙薯蓣（*Disocorea nipponica* Makino.）缠绕茎及根茎。

二、鸢尾科

代表植物：番红花（*Crocus sativus* L.）带花植株。

三、姜科

1. 代表植物 姜（*Zingiber officinales* Rose.）植株、花、果实。

2. 其他药用植物 姜黄（*Curcuma longa* L.）植株。

四、兰科

1. 代表植物 天麻（*Gastrodia elata* Bl.）植株标本及新鲜植株。

2. 其他药用植物 石斛（*Dendrobium nobile* Lindl.）带花植株。

【仪器用品】

放大镜、解剖镜、解剖工具。

【内容与方法】

一、观察解剖下列各代表植物

1. 薯蓣科 薯蓣（附图 18－1）：草质藤本，根状茎叠生，肉质肥厚，圆柱状，生多数须根。茎纤细而长，单叶三角形至三角状卵形，边缘常三浅裂或三深裂，基部耳状膨大，宽心形，叶腋常有珠芽（零余子）。花雌雄异株，花小，排成穗状花序，雄花花被片 6，雄蕊 6，雌花花柱 3，子房下位，柱头 3 裂，蒴果有三棱，成翅状。剖开成熟的果实，可见种子扁圆形，有宽翅。根状茎入药。

2. 鸢尾科 番红花：草本，球茎扁圆球形；叶基生，叶片线形，叶缘反卷。花顶生，有香味；花被管细长，花被裂片 6；雄蕊 3，花药黄色；剥开花被看子房着生部位，子房室数及胎座类型。注意观察花柱的形态，花柱细长，橙红色，顶端 3 深裂，柱头顶端略膨大成漏斗状，边缘有不整齐的锯齿，一侧具裂隙。花柱及柱头入药。

3. 姜科 姜：草本，根状茎肥厚，横生多分枝，折断新鲜根茎，有芳香及辛辣味。叶互生，叶片披针形，无柄，有抱茎的叶鞘。观察花序的着生位置，被以覆瓦状的鳞片，穗状花序球果状；苞片卵形，淡绿色或边缘淡黄色；花冠黄绿色，唇瓣中裂片具紫色条纹及淡黄色斑点，短于花冠裂片，侧裂片卵形；药隔附属体延伸成长喙状。根状茎入药。

4. 兰科 天麻：腐生草本，块茎肥厚，具环节，茎单一，直立，叶鳞片状，膜质，基部呈鞘状抱茎，总状花序顶生；花冠下部壶状，取花一朵仔细观察花的各部分构造特征。蒴果，种子极多，细小，块茎入药。

二、其他药用植物标本的观察

1. 薯蓣科 穿龙薯蓣（穿山龙）（附图 18－2）：缠绕草质，藤本，根状茎横生，圆柱状，外皮成片状脱落。单叶互生，叶掌状心形，边缘有不等大的三角状浅齿，花雌雄异株，穗状花序。蒴果三棱形，每棱翅状。根茎入药。

2. 姜科 姜黄：根状茎断面深黄色至黄红色，折断后有香气。须根末端膨大成淡黄色的块茎。叶基生；注意区分叶鞘、叶片、叶舌。穗状花序自叶鞘内抽出。取花一朵观察，花冠裂片白色；侧生退化雄蕊浅黄色，唇瓣长圆形，花药基部两侧有距。根状茎入药称为姜黄。

3. 兰科 石斛：茎丛生，多节，稍扁；叶互生，长圆形，无柄，具抱茎的鞘。总状花序；花大而艳丽，花萼与花瓣均粉红色，唇瓣近基部中央有一个深紫色大斑块，蕊柱绿色。全草入药。

【实验报告】

1. 写出薯蓣科植物的特征。
2. 写出鸢尾科植物的特征。
3. 绘石斛花的简图，并注明各部分名称。
4. 比较姜科和兰科的主要异同点。

【思考题】

1. 假种皮和种皮有什么不同?
2. 列出中药姜黄、郁金的几种主要植物来源及药用部位。
3. 红花与番红花在来源、形态及功能上有哪些区别?

Experiment 18 Yam Family (Dioscoreaceae), Iris Family (Iridaceae), Ginger Family (Zingiberaceae), Orchid Family (Orchidaceae)

Aim and demand

1. Grasp the main morphological characteristics of the yam family (Dioscoreaceae), iris family (Iridaceae), ginger family (Zingiberaceae), orchid family (Orchidaceae).

2. Recognize the main pharmaceutical plants of the yam family (Dioscoreaceae), iris family (Iridaceae), ginger family (Zingiberaceae), and orchid family (Orchidaceae).

Experiment materials

1. Yam family (Dioscoreaceae)

1.1 Typical plant The branch of vine with flowers and the rhizome of Shuyu (*Dioscorea opposita* Thunb.)

1.2 Other samples of pharmaceutical plants Branches of Chuanlongshuyu (*Dioscorea nipponica* Makino.).

2. Iris family (Iridaceae)

Typical plant: the whole plant with flowers of Fanhonghua (*Crocus sativus* L.).

3. Ginger family (Zingiberaceae)

3.1 Typical plant The total plants, fruits and flowers of Jiang (*Zingiber officinales* Rose.)

3.2 Other samples of pharmaceutical plants Total plant of Jianghuang (*Curcuma longa* L.)

4. Orchid family (Orchidaceae)

4.1 Typical plant Herbarium and fresh plant of Tianma (*Gastrodia elata* Bl.).

4.2 Other samples of pharmaceutical plants The whole plant with flowers of Shihu (*Dendrobium nobile* Lindl.).

Instruments and apliances

Magnifier, anatomical lens, dissector.

Contents and procedures

1. Observe and dissect the typical plants as follows

1.1 Yam family (Dioscoreaceae)

Dioscoreaopposita (Figure 18 - 1): herbaceous twining vines, with tuberous rhizome.

rhizome with many fibres is succulent and columned, the stem tenuity and long; leaves: simple, triangle and triangle ovate, brim 3 partite or dehiscent, basal auriform inflated, cordiform, axil with axillary bud. Flowers: dioecious, small, forming spike, perianth 6 – lobed, stamens 6, distillate flower style 3, inferior ovary, ovary 3 – celled, 3 – winged capsules, dissecting the ripe fruits, to see winged seeds which is oblate; tuberous root is used in making drugs.

1.2 Iris family (Iridaceae)

Crocus sativus: herb, corms, oblate, basal leaf, liner leaves with the brim twisted; terminal flowers with favor, perianth pipe acerose, perianth 6 – lobed, stamens 3. fulvous anther. Pull out the perianth to see the place where ovary is based on, count the number of ovary cells and decide the types of placentation. Remember to observe the shape of the style: acerose, orange, terminal 3 divided; the apex of the style inflate to form a tundish, with irregular serrations on the brim and cranny on lateral. Style and stigma are used in making drugs.

1.3 Ginger family (Zingiberaceae)

Zingiber officinales: herbal, with many branches, snap the fresh rootstocks, there'll be aromatic and pungent. Alternate serrate leaves, sessile, clasping pericladium. Observe the place where the rachis is based on, which with the imbricated elytra, strobiliform spike; bract is ovalliform, light – green or light – yellow on the brim. Flavor – green corolla labial palp median cleavge plane with purple strip and light – yellow spot, shorter than the corolla lobe, lateral lobe is ovalliform. connective accessory substance stretch to a beak shape. Rhizome is used in making drugs.

1.4 Orchid family (Orchidaceae)

Gastrodia elata : saprobic herb, plethoric tuber with ring – joint, stem simple, erect; squamiform leaves, membranous, base with clasp pericladium, terminal racemose inflorescence; the lower corolla urceolate shape, pick a flower to observe the structure feature of every section of it. Capsule, plenty of small seeds. Tuber is used in making drugs.

2. Observe other samples of pharmaceutical plants

2.1 Yam family (Dioscoreaceae)

Dioscorea nipponica (Figure 18 – 2): herbial, wining vine, amphitropic rhizoma, columned, cortex lamellar abscission. Alternate simple leaves, palmate cordate, irregular trigonous dihescent on the brim. Dioecism, spike. 3 – winged capsules. Rhizome is used in making drugs.

2.2 Ginger family (Zingiberaceae)

Curcuma longa: rhizome, section dark – yellows or rufous, cut off with an acromatic smell; fibre terminal inflate to a light – yellow tuber. Basilar leaves; pay attention to distinguish pericladium, blade and ligule. Pike from the pericladium. Pick a flower to observe, corolla section is white, lateral staminode light – yellow, labial palp oblong, there are spurs on the two side of anther base. Rhizome is in drug use, named Jianghuang. Tuber is one of the resource of YuJin a commonly used Traditional Chinese Medicine.

2.3 Orchid family (Orchidaceae)

Dendrobium nobile: collateral stem with plenty of ring - jointed, little compressed; alternate leaves, oblong, sessile, clasping pericladium. racemose inflorescence; flowers huge and flamboyant, pink calyx and pink petals, labial palp near the base central stays a dark - purple spot, green stryle. Total herb is used in making drugs.

Experiment report

1. List the main characteristics of the yam family (Dioscoreaceae).
2. List the main characteristics of the iris family (Iridaceae).
3. Draw the vertical section of flower of *Dendrobium nobile* and note the name of every section .
4. Compare ginger family (Zingiberaceae) with orchid family (Orchidaceae), write the similarities and the differences.

Questions

1. List the main differences between the aril and seed - coat.
2. List the main resources and drug used part of Traditional Chinese Medicine common turmeric (Jianghuang) and aromatic turmeric (Yujin) .
3. List the main different resources, shapes and functions between *Carthamus tinctorius* and *Crocus sativus.*

实验十九　烟草叶片愈伤诱导和植株再生

【目的要求】

1. 掌握一般培养基的配制、灭菌以及无菌操作技术。
2. 学习植物愈伤组织的诱导、调控方法和技术。

【实验材料】

烟草无菌试管苗（由教师提供）：在无菌条件下，将烟草种子用75%乙醇浸泡30s，取出，再用0.1% $HgCl_2$浸泡约3～5min，取出，用无菌水漂洗3～5次，无菌滤纸吸干表面水分，将得到的无菌种子接种在MS培养基上。待种胚苗长出3片叶时继代扩繁备用。

【仪器用品】

1. 仪器　超净工作台，灭菌锅，电炉，不锈钢锅，镊子，剪刀，酒精灯，100ml广口三角烧瓶40个，量筒（100ml，1个），移液管（5ml，2支），pH计（或pH精密试纸5.4～7.0），架盘天平（200g），牛皮纸，硫酸纸，细绳，培养皿。

2. 试剂　MS培养基母液①、②、③、④，激素母液：6-BA（0.5mg/ml），NAA（0.5mg/ml），蔗糖，琼脂粉，NaOH溶液（1mol/L），HCl溶液（1mol/L）。

【内容与方法】

一、培养基的配制

愈伤诱导培养基的配制：配方为 MS + 蔗糖（20g/L） + 琼脂（7g/L） + BA（2.0mg/L） + NAA（2.0 mg/L）。欲配制 1L 培养基，先用量筒量取 MS 培养基母液①100ml、②10ml、③10ml、④10ml，再用移液管精密量取激素母液 BA 4ml，NAA 4ml，混合上述 6 种母液，加入蔗糖 20g，搅拌使溶化，将混合液在量筒中用蒸馏水定容至 1000ml，混合均匀，用 pH 计或 pH 精密试纸测定混合液的 pH，用 1mol/L 的氢氧化钠或盐酸溶液调至 5.7 左右。将混合液置于不锈钢锅中，加热，加入 7g 琼脂，不断搅拌，保持混合液沸腾状态 2min 左右，由混浊变完全澄清时停止加热，在量筒中再次定容至 1L。分装至 20 个锥形瓶中，用牛皮纸和硫酸纸做好的纸盖盖好，再用细绳绑住瓶口，放在灭菌锅内，按照操作规程灭菌，在 121℃条件下保持 15min。待灭菌锅温度压力下降到 0.5MPa 时打开放气阀，使压力降到零，取出三角烧瓶，置超净工作台上备用。

二、生苗培养基配制

培养基配方：MS + 蔗糖（20g/L） + 琼脂（7g/L） + BA（2.0mg/L） + NAA（0.5 mg/L）。配制、灭菌方法同上。

三、烟草叶片愈伤组织诱导

在超净工作台上取上述烟草无菌试管苗幼嫩叶片 1 ~ 2 片于无菌培养皿中，用无菌剪刀将叶片剪成 $2mm^2$ 左右的小片并接种到愈伤诱导培养基中，每瓶接种 5 片，共接种 20 瓶，依次编号为 1 ~ 20 号。将 1 ~ 20 号均置于 24℃条件下黑暗培养 1 周，然后在同样的温度下将 1 ~ 10 号置于光照条件下培养，11 ~ 20 号置于黑暗条件下培养。

四、烟草植株的再生

在超净工作台上，将上面实验中得到的愈伤组织按来源类型（光照条件下还是黑暗条件下）转到生苗培养基中，接种 20 瓶，做好标示（如：光 1，光 2……暗 1，暗 2……）。每瓶 5 块，20 ~ 22℃条件下连续光照培养。

【实验报告】

1. 记录组织培养的一般程序以及需要注意的问题。
2. 描述光照培养和黑暗培养两种条件下每瓶中烟草叶片愈伤组织的形态、颜色、质地等特征，以及两种条件下诱导产生的愈伤组织在相同条件下分化成苗能力的差异。
3. 计算光照培养和黑暗培养两种条件下烟草叶片愈伤组织的诱导率。
4. 计算两种条件下得到的愈伤组织在相同条件下的分化率。

【思考题】

根据实验结果，讨论不同的培养条件对外植体愈伤组织的诱导、愈伤组织再生植株能力的影响。

Experiment 19 Callus Induction and Plant Regeneration From Leaf Explants of Tobacco

Aim and demand

1. Gain the technology of preparing and sterilizing of common media and the operating without infection.

2. Learn the methods of the callus inducing and conditions controlling.

Experiment materials

Asepsis young plants of tobacco (*Nicotiana tabacum* L.) in test tubes (Provided by teacher): We can obtain the asepsis seeds of tobacco after the following operations. Marinate the seeds of tobacco inethanol (75%) for about 30 seconds in sterilized environment. Take out the seeds, marinate them in the solution of $HgCl_2$ (0.1%) for about 3 – 5 minutes. After the operations above, wash the seeds for 3 – 5 times with sterilized distilled water, then sip up the water on the surface of the seeds with asepsis filter paper. Transplant the asepsis seeds in sterilized MS medium in the conical flasks. When the plants from the embryo have 3 leaves, we can continue their propagation for latter use.

Instruments and aplliances

1. Instruments

Clean – operating board, pressure cooker, micro – wave oven, tweezers, scissors, alcohol burner, conical flasks (100ml, 40), graduated flask (100ml, 1), liquid – shift tube (5ml, 2), pH – meter or pH indicator paper (5.4 – 6.8), balance (200g), brown paper, vitriol paper, culture utensil.

2. Reagents

MS media mother solution ①, ②, ③, ④; hormone mother solution: 6 – BA (6 – benzylaminopurine) (0.5mg/ml); NAA (naphthylacetic acid) (0.5mg/ml); cane sugar; agar powder; NaOH solution (1mol/L); HCl solution (1mol/L).

Contents and procedures

1. Media Preparation

1.1 The components of the media

MS + sucrose (20g/L) + agar powder (7g/L) + BA (2.0mg/L) + NAA (2.0 mg/L).

1.2 Media preparation

In order to prepare 1L medium, firstly get the following stocksolution of told volume respectively with measuring cylinder and liquid – shift tube: MS①100ml, ②10ml, ③10ml, ④10ml; BA 4ml, NAA 4ml, then mix the six kinds of liquids of requested. Add the sucrose (20g) to the mixture and stir until it melts completely. (Don't add the agar (gelling agent) at this stage because it gums things up when adjusting the pH). Bring the volume to 1L with distilled water and mix well by stirring. Adjust the pH to pH 5.7 using 1mol/L NaOH or 1mol/L HCl (carefully, by drop) and pH indicator paper (3.5 - 6.8), or a pH meter. (If you don't have a pH meter, the chemistry teacher might.). Now add 7g agar. Heat and stir until the medium is clear completely by maintaining the boiling state for about 2min (The clarity tells you that the agar has melted.). Then make sure the volume of the media is 1000ml again in the measuring cylinder and dispense the media into 20 conical flasks.

Cover the mouth of flasks dispensed with media made of brown paper and vitriol paper. Then fasten the covers with strings. Place the flasks into autoclave. According to the autoclave instructions, keep for 15min at 121℃. Open the deflation valve until the pressure of the cooker is less than 0.5MPa. Take out the flasks place them on clean bench.

2. The component of the media of regenerating

MS + cane sugar (20g/L) + agar powder (7g/L) + BA (2.0mg/L) + NAA (0.5 mg/L).

The methods of preparing and sterilizing are almost the same as the one mentioned above.

3. The callus Induction from the leaves of tobacco

Choose freshasepsis leaves as explants. These explants were cut into pieces about $2mm^2$, and then inoculated into conical flasks and five pieces for each dispensed with the media of callus inducing in asepsis environment on the table of super - clean - operating board. Number the inoculated conical flasks 1 - 20. Put these flasks in darkness and at the temperature of 24℃ for a week's time. Then put the flasks numbered 1 - 10 in lightness by cool white fluorescent lamps at the same temperature, and let the others put still in darkness.

4. The regeneration of tobacco plants

Inoculate the callus into media for regenerating obtained from the experiment above on the table of super - clean - operating board. Five pieces for each flask. Then mark the flask inoculatedrespectively according to its culturing conditions (darkness or lightness). Put all of the flasks under continuous lightness by cool white fluorescent lamps under the temperature of 20 - 22℃.

Experiment report

1. Note the common process of tissue culture and problems to notice.
2. Describe the shape, colour and texture of the callus and the difference between theircapabilities of differentiation in same conditions cultivated both in lightness and darkness.
3. Account the inducement rate of the callus from tobacco leaves cultivated under lightness and darkness.
4. Account the differentiation rate of the callus under the same cultivated conditions.

Questions

According to the result of your experiment, please discuss the effect of different conditions to the callus induction of the explants and the capability of differentiating or regenerating of callus.

附录
Appendix

一、常用试剂的配制

1. 清洁剂 取乙醚7ml和无水乙醇3ml混合，装入滴瓶备用，用于擦拭显微镜镜头上的油迹和污垢等（瓶口必须塞紧，以免挥发）。

2. F. A. A 固定液（又称万能固定剂） 福尔马林（38%甲醛）5ml，冰醋酸5ml，70%乙醇90ml。幼嫩材料用50%乙醇代替70%乙醇，可防止材料收缩；还可以加入甘油5ml，以防止蒸发和材料变硬。此液兼可作为保存剂使用。

3. 稀甘油 取甘油33ml，加蒸馏水稀释成100ml，再加樟脑一小块或液化苯酚1滴，即得。

稀甘油能使细胞稍透明及溶解某些水溶性细胞后含物，并使材料保持湿润和软化，常和水合氯醛试剂同用作临时封藏剂，可防止水合氯醛晶体析出。

4. 水合氯醛试剂 取水合氯醛50g，加蒸馏水15ml，与甘油10ml使其溶解，即得。本试液能迅速透入组织，使干燥而收缩的细胞膨胀，细胞组织透明清晰，并能溶解淀粉粒、树脂、蛋白质和挥发油等。

5. 苯三酚试液 取间苯三酚1g，加90%乙醇5ml溶解后，加甘油5ml，摇匀，即得。用于鉴别木质化细胞壁，应用时先加1~2滴于检体，约1min后，加盐酸1滴，木质化细胞壁因木质化程度不同而显红色或紫色。

6. 稀碘液 取碘化钾1g溶于100ml蒸馏水中，再加碘0.3g，储于棕色瓶中。稀碘液可使淀粉粒显蓝色，糊粉粒显黄色。

7. 苏丹Ⅲ试液 取苏丹Ⅲ0.01g，加90%乙醇5ml溶解后，加甘油5ml，摇匀，储于棕色玻璃瓶中，保存期为2个月。本试液能使角质化和木栓化细胞壁显红色或橙红色，使脂肪油、挥发油或树脂显橙红色、红色或紫红色。

8. 钌红试液 取10%醋酸钠溶液1~2ml，加钌红适量使成酒红色，即得。本试液应临用时新制，可使黏液显红色。

9. α-萘酚试液 取α-萘酚1.5g溶于95%乙醇10ml，即得。应用时滴加本试液，1~2min后再滴加80%硫酸2滴，可使菊糖显紫色。

10. 氯化锌碘试液 取氯化锌20g，溶于85ml蒸馏水后，滴加碘的碘化钾溶液（碘化钾3g，碘1.5g，水60ml），不断振摇至饱和，至没有碘的沉淀出现为止，至棕色瓶内保存。本试液可使纤维素细胞壁显蓝色和紫色。

11. 番红染液 番红是一种碱性染料，可使木质化、木栓化、角质化的细胞壁及细胞核中的染色质和染色体染成红色。在植物组织制片中常与固绿配染。常用配方有下列两种。

（1）番红水液　取番红0.1g，溶于100ml蒸馏水中，过滤后，即得。

（2）番红酒液　取番红0.5g或1g，溶于50%乙醇100ml中，过滤后，即得。

12. 固绿染液　固绿是一种酸性染料，可使纤维素的细胞壁和细胞质染成绿色。在植物组织制片中常与番红配染。常用固绿酒液，即取固绿0.1g，溶于95%乙醇100ml中，过滤后使用。

13. 洗液　取工业用重铬酸钾（$K_2Cr_2O_7$）8～10g于100ml清水中，加热使溶解，待冷却后，慢慢加入工业用浓硫酸（H_2SO_4）100ml，即得。储藏于玻璃容器中。

14. IAA（0.5mg/ml）　先用少量95%乙醇使50mg IAA粉末充分溶解，再加蒸馏水定容至100ml。

15. 2，4－D：先用少量1mol/L的NaOH溶液充分溶解，然后缓慢加入蒸馏水定容至需要体积。

16. 细胞分裂素类　KT和BA等细胞分裂素类物质均溶于稀盐酸，应先用少量1mol/L的HCl溶解后再稀释至需要浓度。

17. 赤霉素　赤霉素的水溶液稳定性较差，一般用95%的乙醇配制成5～10mg/ml的母液低温保存，使用时再稀释。

Ⅰ. The Preparation and Use of Frequently Used Regents

1. Tergent　Get ether 7ml mixed with ethanol 3ml, fill the mixture in a bottle for latter use. this kind of detergent is for the use of wiping the oil and dirt on the lens of the microscope.（note: the bottle must be corked tighely in order to prevent the detergent from volatilising）

2. F. A. A reagent for fixing　Also called universal reagent for fixing. Get formaldehyde solution（38%）5ml mixed with glacial acetic acid 5ml and ethanol solution（70%）90ml. if the material delt with is rather tender, please use ethanol 50% in stead of ethanol 70% in order to protect the material from contracting . besides we can add glycerine 5ml in the prepared reagent for the use of preventing volatilising and the material from turning harder. the reagent also can be used for preserving.

3. Thin glycerine　Get glycerine 30ml diluted to 100ml with distilled water , then add a drop of liquefied phenol or a piece of camphor to the glycerine solution . The thin glycerine could make cells seem more transparent and dissolve some hydrosoluble ergastic substance, soften the materials and keep them being moist . The glycerine is usually used with chloral hydrate as a temporary sealing reagent and can prevent the crystal of chloral hydrate from separating out.

4. Chloral hydrate reagent　Get chloral hydrate 50g mixed homogeneously with distilled water 15ml and glycerine 10ml. This reagent can seep into the material quickly, make the dry and contracted cells swell, the tissue of cells transparent and distinct. besides , the reagent can dissolve substances such as starch grains, resin, protein, and essential oil and so on.

5. Trihydroxy－phenole reagent　Get trihydroxy－phenole 1g dissolved in alcohol solution (90%) 5ml, then mixed homogeneously with glycerine 5ml. The reagent is used to identify xylinus cell wall. Method of preparing: we can add several drops of the reagent to the material delt with, then add a drop of hydrochloric acid to it after about one minute. because of the difference of the degree of xylinus of cell wall, the cell wall will turn red to purplishred.

6. Thin iodine solution Get potassium iodide 1g dissolved in distilled water 100ml, then add iodine 0.3g to the solution. This reagent can make the starch grain turn blue, and the alcurone grain turn yellow. Please preserve the solution in a brown bottle.

7. Sudan Ⅲ regent Get Sultan Ⅲ 0.01g dissolved in alcohol solution (90%) 5ml, and mixed homogeneously with glycerine 5ml. Please preserve the solution in a brown bottle. The regent can be effective within two - month's time long. The regent can make the cell wall turn red that is cutined or phellemed and fixed oil, essential oil, resin turn orange red, red, or purplish red.

8. Ruthwnium red regent Get solution (10%) of sodium acetic 1 ~ 2ml. Add an appropriate amount of Ruthwnium red, and make the solution be wine red. The regent can make the mucilage turn red. The regent should be prepared only when you are using it.

9. α - naphthol regent Get α - naphthol 1.5g dissolved in alcohol solution (95%). When using the solution, add on the material two drops of sulphuric acid solution (80%) after adding the regent. The operation will make the inulins turn purple within one or two minutes.

10. $ZnICl_3$ regent Get Zine chloride 20g dissolved in distilled water 85ml, then add Iodine's solution of Potassium Iodine (KI: 3g, I_2: 1g, distilled water: 60ml). Shake the solution in a bottle until it is saturate and there are no sediment of I_2. Please preserve the solution in brown bottle. The regent can make the cellulose - cell - wall turn blue or purple.

11. Safranine dye solution Safranine is a kind of basic dye. It can make chromatin, chromosome in the karyon of cells and the cell wall that is lignified, corking over, keratinizated turn red. It is usually used with green - fixed dye in the slice - making of bortany's tissue experiment. The methods of it's preparing are the two followings:

(1) Safranine dissolved in water Get Safranine 0.1g dissolved in distilled water 100ml, then get the solution filtered.

(2) Safranine dissolved in alcohol Get Safranine 0.5g dissolved in alcohol solution (50%) 100ml, then get the solution filtered.

12. Green - fixed dye solution Green - fixed dye is a kind of acidic dye, it can make cytoplasm or cell wall that contains cellulose turn green. It is usually used with Safranine in the slice - making of plant's tissue experiment. It's solution in alcohol is used more frequently. The alcohol solution can be prepared as follows: get Green - fixed dye 0.1g dissolved in alcohol solution (95%) 100ml, and then filter the solution.

13. Lotion Get Industrial $K_2Cr_2O_7$ 8 - 10g, and add it to water 100ml, then heat the solution and make sure the substance dissolve completely; and then add slowly the Industrial concentrated sulphuric acid 100ml to the solution until the solution is cold after heated. Please preserve the lotion in a glass container.

14. IAA（0.5mg/ml） Get 50 mg of IAA powder and make it dissolve completely with alcohol（95%）, and then add some distilled water to 100ml.

15. 2, 4-D Dissolve the powder completely with NaOH solution（1mol/L）, then add the distilled water slowly to the required volume.

16. KT and BA Dissolve the powder completely with a little HCl solution（1mol/L）, then dilute the solution to the required volume.

17. Gibberellin Dissolve the powder with alcohol（95%）, dilute the solution to the concentration of 5-10mg/ml and store the solution in low temperature. Dilute again the solution to the required concentration when using it.

二、《中国植物志》卷册索引

《中国植物志》是由80卷125册组成，为了便于查找和使用，现附上卷册索引。数字为卷号，括号内数字为册号。

（一）《中国植物志》各卷内容

1 总论

2，3（1）（2），4（1）（2），5，6（1）（2）蕨类植物

7 裸子植物

8 香蒲科、露兜树科、黑三菱科、眼子菜科、茨藻科、水蕹科、泽泻科、花蔺科、水鳖科、霉草科、冰沼草科（芝菜科）

9（1）（2）（3），10（1）（2）禾本科

11，12 莎草科

13（1）棕榈科

13（2）天南星科、浮萍科

13（3）须叶藤科、帚灯草科、刺鳞草科、黄眼草科、谷精草科、凤梨科、鸭跖草科、雨久花科、田葱科、灯芯草科、百部科

14，15 百合科

16（1）石蒜科、蒟蒻薯科、薯蓣科、鸢尾科

16（2）芭蕉科、姜科、美人蕉科、竹芋科、水玉簪科

17，18，19 兰科

20（1）木麻黄科、三白草科、胡椒科、金粟兰科

20（2）杨柳科

21 杨梅科、胡桃科、桦木科

22 壳斗科（山毛榉科）、榆科、马尾树科

23（1）桑科

23（2）荨麻科

24 川苔草科、山龙眼科、铁青树科、山柚仔科、檀香科、桑寄生科、马兜铃科、大花草科、蛇菰科

25（1）蓼科

25（2）藜科、苋科

26 紫茉莉科、商陆科、番杏科、马齿苋科、落葵科、石竹科

27 睡莲科、金鱼藻科、昆栏树科、领春木科、连香树科、毛茛科

28 毛茛科

29 木通科、小檗科

30（1）防己科、木兰科

30（2）蜡梅科、番荔枝科、肉豆蔻科

31 樟科、莲叶桐科

32 罂粟科、山柑科

33 十字花科

34（1）木犀草科、辣木科、伯乐树科（钟萼木科）、猪笼草科、茅膏菜科、景天科

34（2），35（1）虎耳草科

35（2）海桐花科、金缕梅科、杜仲科、悬铃木科

36，37 蔷薇科

38 蔷薇科、牛栓藤科

39，40，41，42（1）（2）豆科

43（1）攀打科、酢浆草科、牻牛儿苗科、旱金莲科、亚麻科、古柯科、蒺藜科

43（2）芸香科

43（3）苦木科、橄榄科、楝科、金虎尾科、远志科、毒鼠子科

44（1）（2）（3）（4）大戟科

45（1）虎皮楠科、水马齿科、黄杨科、岩高兰科、马桑科、漆树科、五列木科

45（2）冬青科

45（3）卫矛科

46 翅子藤科、刺茉莉科、省沽油科、茶茱萸科、槭树科、七叶树科

47（1）无患子科、清风藤科

47（2）凤仙花科

48（1）鼠李科

48（2）葡萄科

49（1）杜英科、椴树科

49（2）锦葵科、木棉科、梧桐科、五桠果科、猕猴桃科、金连木科

49（3），50（1）山茶科

50（2）藤黄科（金丝桃科）、龙脑香科、沟繁缕科、瓣鳞花科、柽柳科、半日花科、红木科

51 堇菜科

52（1）大风子科、旌节花科、西番莲科、番木瓜科、四数木科、秋海棠科、钩枝藤科、仙人掌科、瑞香科

52（2）胡颓子科、千屈菜科、海桑科、隐翼科、石榴科（番石榴科）、玉蕊科、红树科、蓝果树科（珙桐科）、八角枫科

53（1）使君子科、桃金娘科、野牡丹科

53（2）菱科、柳叶菜科、小二仙草科、杉叶藻科、假繁缕科、锁阳科

54 五加科

55（1）（2）（3）伞形科

56 山茱萸科（四照花科）、岩梅科、桤叶树科（山柳科）、鹿蹄草科

57（1）（2）（3）杜鹃花科

58 紫金牛科

59（1）（2）报春花科

60（1）白花丹科、山榄科、柿科

60（2）山矾科、安息香科（野茉莉科）

61 木犀科、马钱科

62 龙胆科

63 夹竹桃科、萝藦科

64（1）旋花科、花荵科、田基麻科

64（2）紫草科

65（1）马鞭草科

65（2），66 唇形科

67（1）茄科

67（2），68 玄参科

69 紫葳科、胡麻科、角胡麻科、列当科、苦苣苔科、狸藻科

70 爵床科、苦槛蓝科、透骨草科、车前科

71（1）（2）（3）茜草科

72 忍冬科

73（1）五福花科、败酱科、川续断科、葫芦科

73（2）桔梗科、草海桐科、花柱草科

74～80（1）（2）菊科

（二）《中国植物志》被子植物各科拉丁文科名卷册索引

（按科名首字母顺序排列）

A

Acanthaceae（70）

Aceraceae（46）

Actinidiaceae（49（2））

Adoxaceae（73（1））

Aizoaceae（26）

Alangiaceae（52（2））

Amaranthaceae（25（2））

Amaryllidaceae（16（1））

Anacardiaceae（45（1））

Ancistrocladaceae（52（1））

Annonaceae（30（2））

Aponogetonaceae（8）

Araceae（13（2））

Araliaceae（54）

Aristolochiaceae（24）

Asclepiadaceae (63)

B

Balanophoraceae (24)
Balsaminaceae (47 (2))
Basellaceae (26)
Begoniaceae (52 (1))
Berberidaceae (29)
Betulaceae (21)
Bignoniaceae (69)
Bixaceae (50 (2))
Bombacaceae (49 (2))
Boraginaceae (64 (2))
Bretschneideraceae (34 (1))
Bromeliaceae (13 (3))
Burmanniaceae (16 (2))
Burseraceae (43 (3))
Butomaceae (8)
Buxaceae (45 (1))

C

Cactaceae (52 (1))
Callitrichaceae (45 (1))
Calycanthaceae (30 (2))
Campanulaceae (73 (2))
Cannaceae (16 (2))
Capparaceae (32)
Caprifoliaceae (72)
Caricaceae (52 (1))
Caryophyllaceae (26)
Casuarinaceae (20 (1))
Celastraceae (45 (3))
Centrolepidaceae (13 (3))
Ceratophyllaceae (27)
Cercidiphyllaceae (27)
Chenopodiaceae (25 (2))
Chloranthaceae (20 (1))
Cistaceae (50 (2))
Clethraceae (56)
Combretaceae (53 (1))
Commelinaceae (13 (3))
Compositae (74 ~ 80 (1), (2))
Connaraceae (38)
Convolvulaceae (64 (1))
Coriariaceae (45 (1))
Cornaceae (56)
Crassulaceae (34 (1))
Cruciferae (33)
Crypteroniaceae (52 (2))
Cucurbitaceae (73 (1))
Cynomoriaceae (53 (2))
Cyperaceae (11, 12)

D

Daphniphyllaceae (45 (1))
Diapensiaceae (56)
Dichapetalaceae (43 (3))
Dilleniaceae (49 (2))
Dioscoreaceae (16 (1))
Dipsacaceae (73 (1))
Dipterocarpaceae (50 (2))
Droseraceae (34 (1))

E

Ebenaceae (60 (1))
Elaeagnaceae (52 (2))
Elaeocarpaceae (49 (1))
Elatinaceae (50 (2))
Empetraceae (45 (1))
Ericaceae (57 (1), (2), (3))
Eriocaulaceae (13 (3))
Erythroxylaceae (43 (1))
Eucommiaceae (35 (2))
Euphorbiaceae (44 (1), (2), (3))
Eupteleaceae (27)

F

Fagaceae (22)
Flacourtiaceae (52 (1))
Flagellariaceae (13 (3))
Frankeniaceae (50 (2))

G

Gentianaceae（62）
Geraniaceae（43（1））
Gesneriaceae（69）
Goodeniaceae（73（2））
Gramineae（9～10）
Guttiferae（50（2））

H

Haloragidaceae（53（2））
Hamamelidaceae（35（2））
Hernandiaceae（31）
Hippocastanaceae（46）
Hippuridaceae（53（2））
Hydrocharitaceae（8）
Hydrophyllaceae（64（1））

I

Icacinaceae（46）
Iridaceae（16（1））

J

Juglandaceae（21）
Juncaceae（13（3））

L

Labiatae（65（2），66）
Lardizabalaceae（29）
Lauraceae（31）
Lecythidaceae（52（2））
Leguminosae（39～42）
Lemnaceae（13（2））
Lentibulariaceae（69）
Liliaceae（14，15）
Linaceae（43（1））
Loganiaceae（61）
Loranthaceae（24）

M

Magnoliaceae（30（1））
Malpighiaceae（43（3））
Malvaceae（49（2））
Marantaceae（16（2））
Martyniaceae（69）
Melastomataceae（53（1））
Menispermaceae（30（1））
Moraceae（23（1））
Moringaceae（34（1））
Musaceae（16（2））
Myoporaceae（70）
Myricaceae（21）
Myristicaceae（30（2））
Myrsinaceae（58）
Myrtaceae（53（1））

N

Najadaceae（8）
Nepenthaceae（34（1））
Nyctaginaceae（26）
Nymphaeaceae（27）
Nyssaceae（52（2））

O

Ochnaceae（49（2））
Olacaceae（24）
Oleaceae（61）
Onagraceae（53（2））
Opiliaceae（24）
Orchidaceae（17～19）
Orobanchaceae（69）
Oxalidaceae（43（1））

P

Palmae（13（1））
Pandanaceae（8）
Papaveraceae（32）
Passifloraceae（52（1））
Pedaliaceae（69）
Pentaphylacaceae（45（1））
Philydraceae（13（3））
Phrymaceae（70）
Phytolaccaceae（26）
Piperaceae（20（1））
Pittosporaceae（35（2））

Plantaginaceae (70)
Platanaceae (35 (2))
Plumbaginaceae (60 (1))
Podostemaceae (24)
Polemoniaceae (64 (1))
Polygalaceae (43 (3))
Polygonaceae (25 (1))
Pontederiaceae (13 (3))
Portulacaceae (26)
Potamogetonaceae (8)
Primulaceae (59 (1), (2))
Proteaceae (24)
Punicaceae (52 (2))
Pyrolaceae (56)

R

Rafflesiaceae (24)
Ranunculaceae (27, 28)
Resedaceae (34 (1))
Restionaceae (13 (3))
Rhamnaceae (48 (1))
Rhizophoraceae (52 (2))
Rhoipteleaceae (22)
Rosaceae (36 – 38)
Rubiaceae (71 (1), (2))
Rutaceae (43 (2))

S

Sabiaceae (47 (1))
Salicaceae (20 (2))
Salvadoraceae (46)
Santalaceae (24)
Sapindaceae (47 (1))
Sapotaceae (60 (1))
Saururaceae (20 (1))
Saxifragaceae (34 (2), 35 (1))
Scheuchzeriaceae (8)
Simaroubaceae (43 (3))
Solanaceae (67 (1))
Sonneratiaceae (52 (2))
Stachyuraceae (52 (1))
Staphyleaceae (46)
Stemonaceae (13 (3))
Sterculiaceae (49 (2))
Stylidiaceae (73 (2))
Styracaceae (60 (2))
Symplocaceae (60 (2))

T

Taccaceae (16 (1))
Tamaricaceae (50 (2))
Theaceae (49 (3), 50 (1))
Theligonaceae (53 (2))
Thymelaeaceae (52 (1))
Tiliaceae (49 (1))
Trapaceae (53 (2))
Triuridaceae (8)
Trochodendraceae (27)
Tropaeolaceae (43 (1))
Typhaceae (8)

U

Ulmaceae (22)
Umbelliferae (55 (1), (2), (3))
Urticaceae (23 (2))

V

Valerianaceae (73 (1))
Verbenaceae (65 (1))
Violaceae (51)
Vitaceae (48 (2))

X

Xyridaceae (13 (3))

Z

Zingiberaceae (16 (2))
Zygophyllaceae (43 (1))

附图

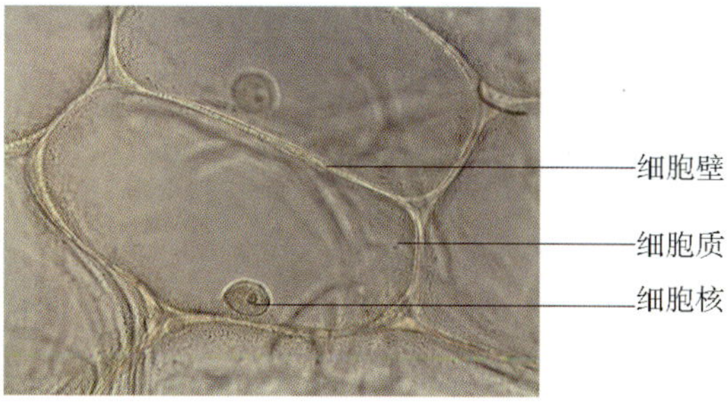

附图 1-1 洋葱鳞茎内表皮细胞

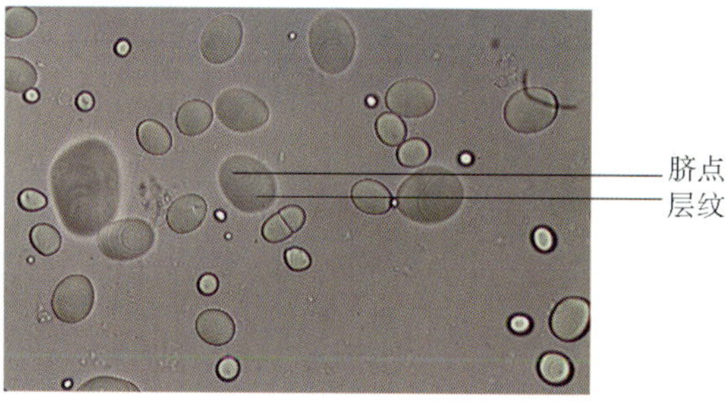

附图 1-2 马铃薯块茎淀粉粒

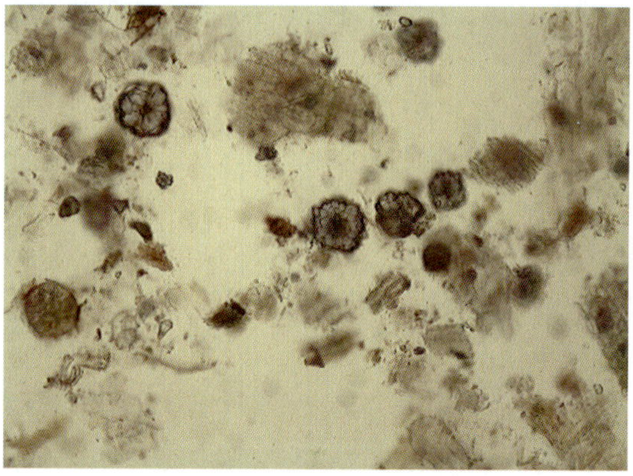

附图 1-3 大黄根茎簇晶

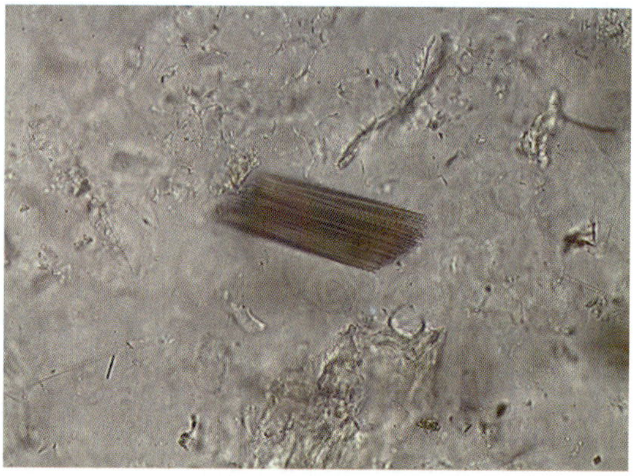

附图 1-4 半夏根茎针晶

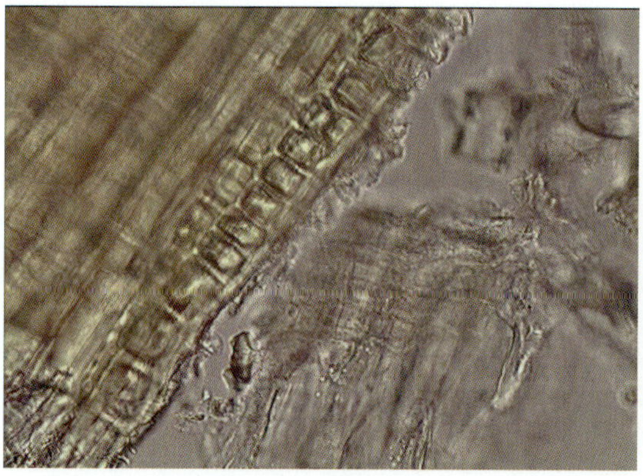

附图 1-5 黄柏皮方晶

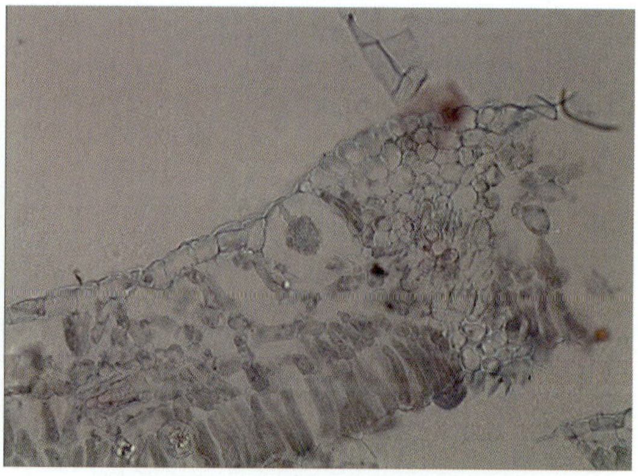

附图 1-4 无花果叶碳酸钙晶体

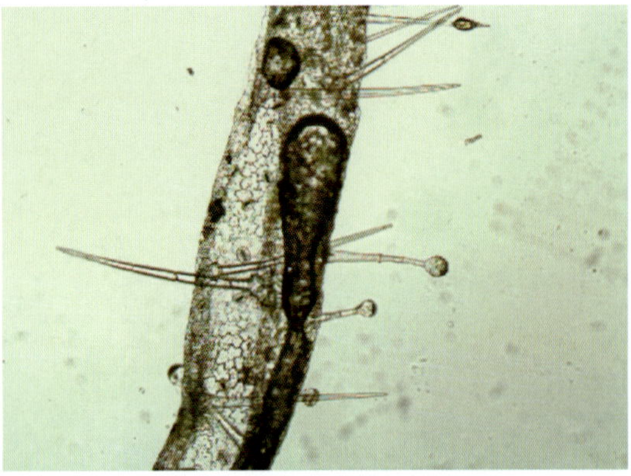

附图 2-1 天竺葵叶腺毛、非腺毛

附图 2-2 薄荷叶腺鳞

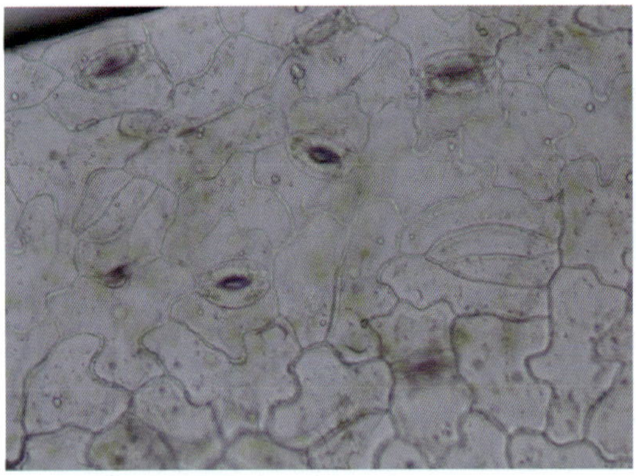

附图 2-3 常山叶平轴式气孔

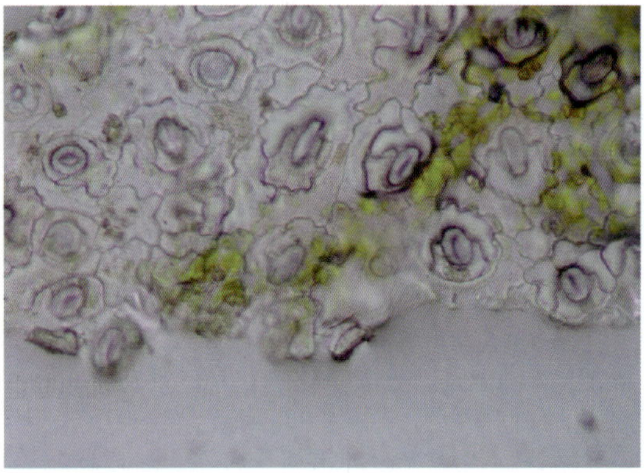

附图 2-4 薄荷叶直轴式气孔

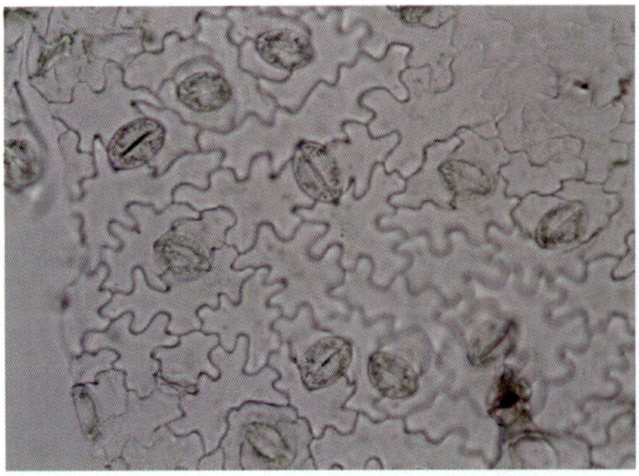

附图 2-5 曼陀罗叶不等式气孔

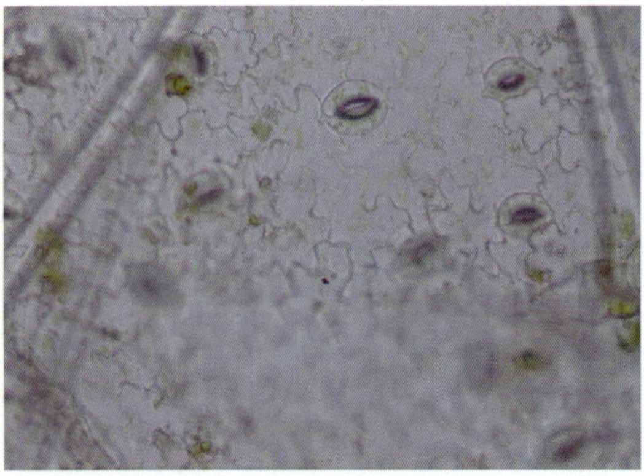

附图 2-6 天竺葵叶不定式气孔

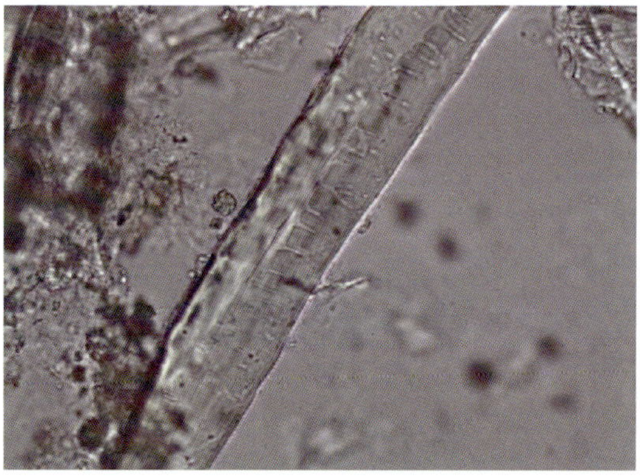

附图 2–7 白藓皮根纤维

附图 3–1 松茎管胞

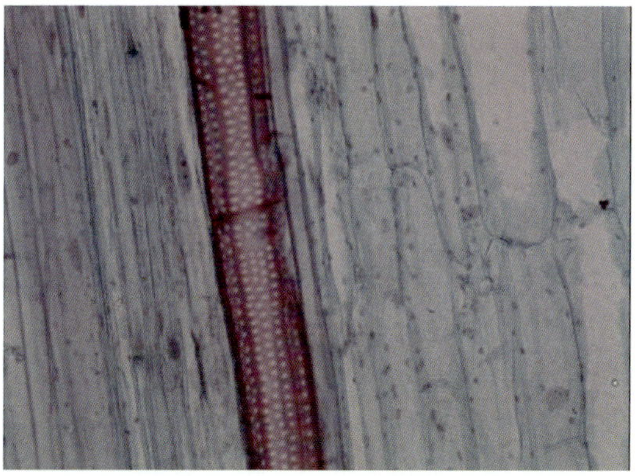

附图 3–2 向日葵茎网纹导管

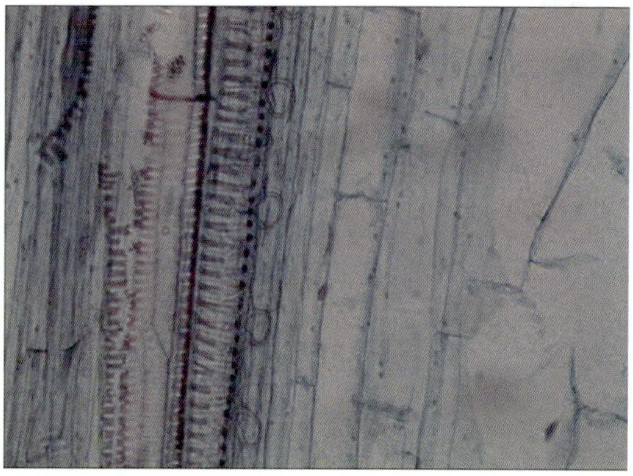

附图 3–3 向日葵茎环纹、螺纹导管

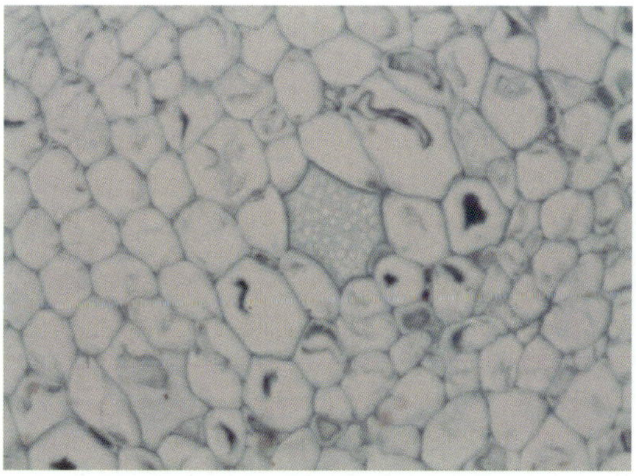

附图 3–4 南瓜茎筛管、筛板

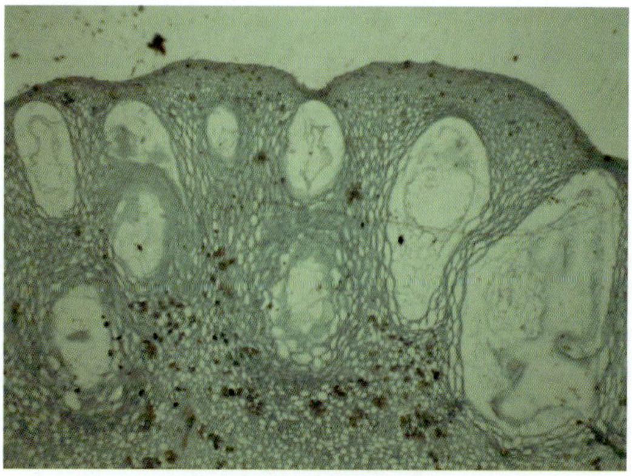

附图 3–5 陈皮油室

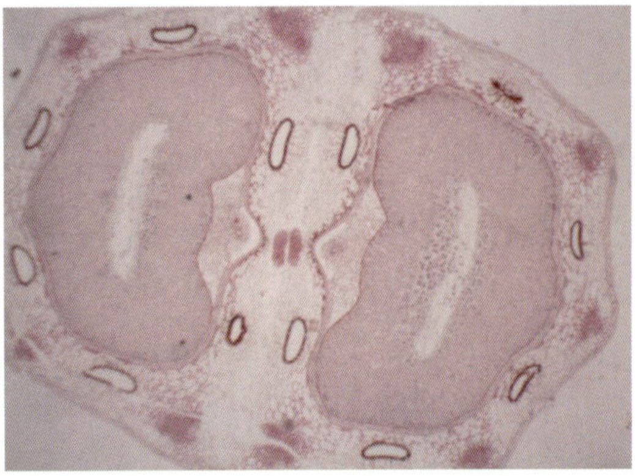

附图 3-6 小茴香果实油管

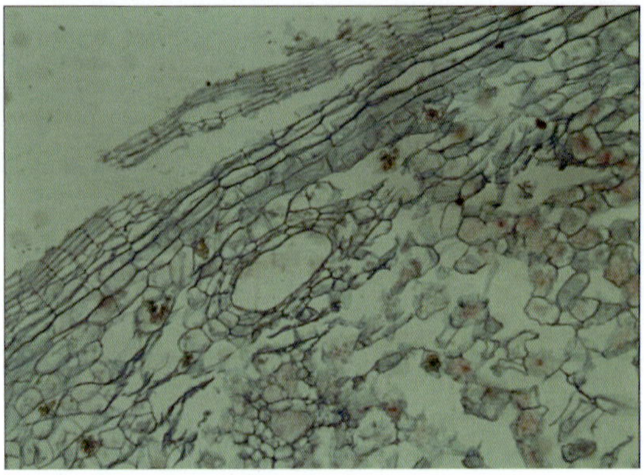

附图 3-7 松茎树脂道

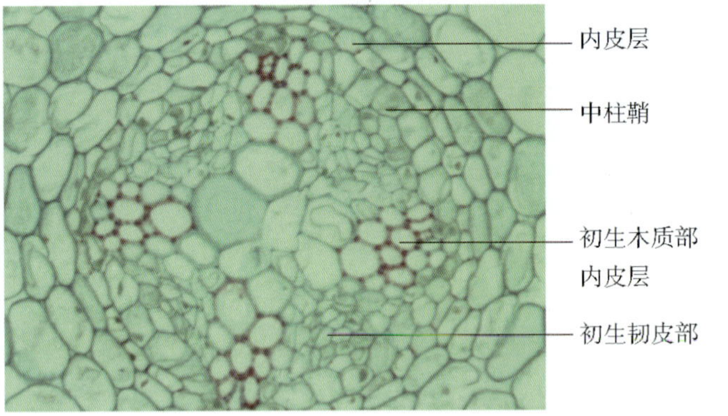

附图 4-1 毛茛根的横切片

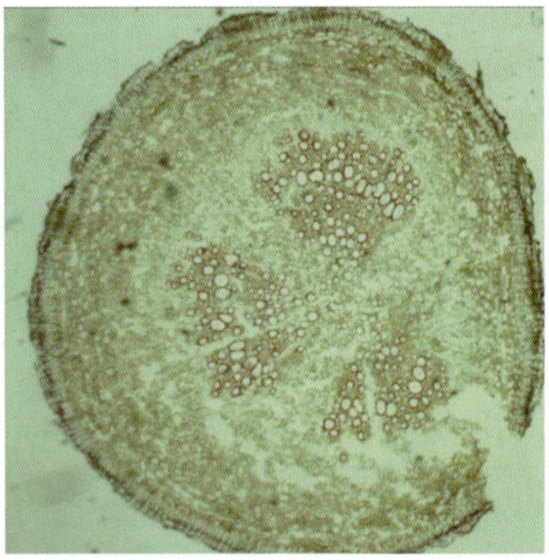

附图 4-2 芍药根横切片

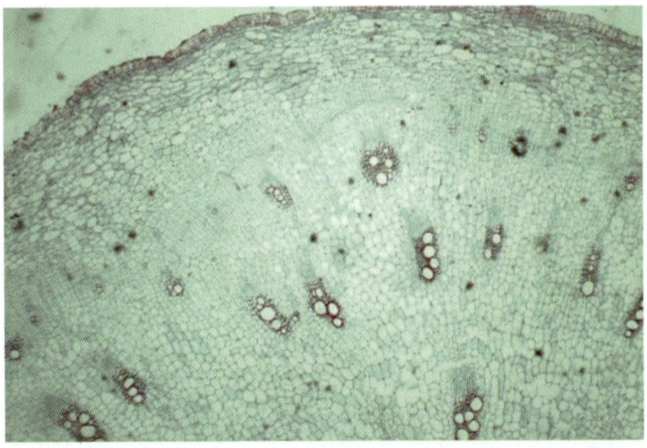

附图 4-3 牛膝根横切片

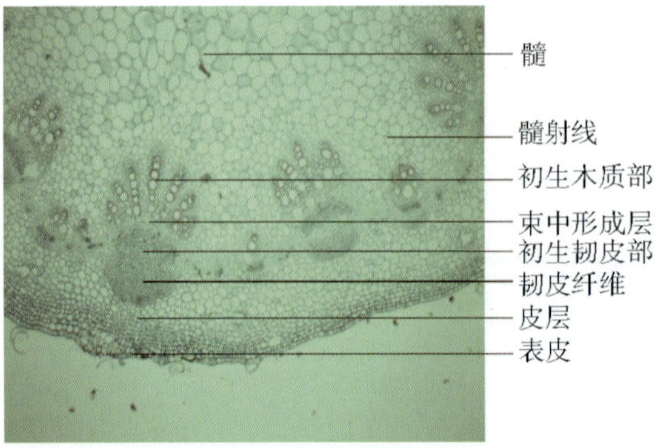

附图 5-1 向日葵幼茎横切片

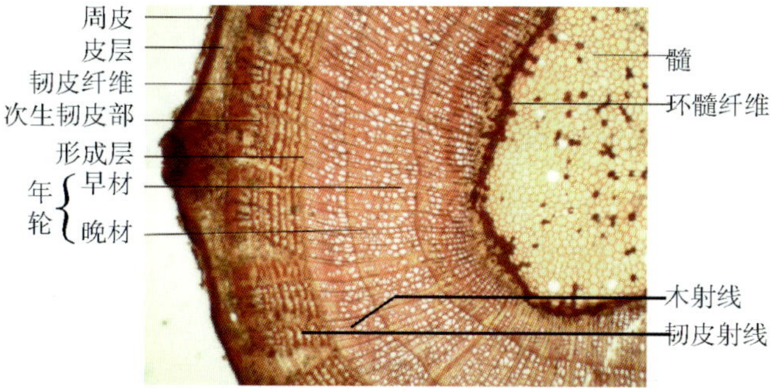

附图 5–2　椴树茎的横切片

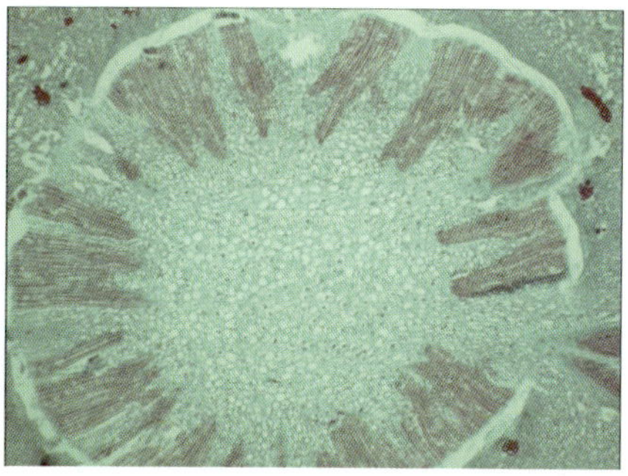

附图 6–1　黄连根茎横切片

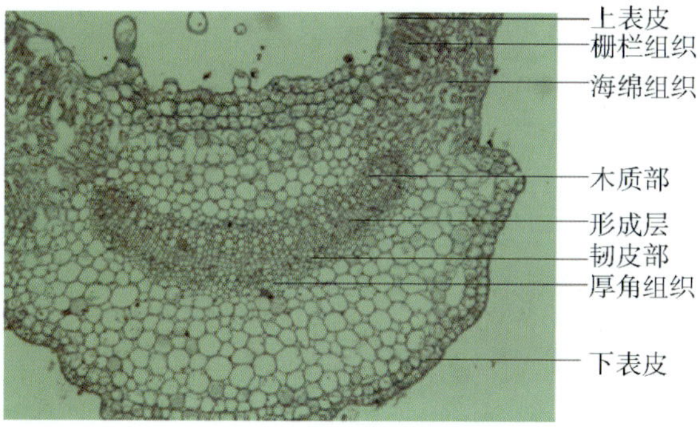

附图 7–1　薄荷叶的横切片

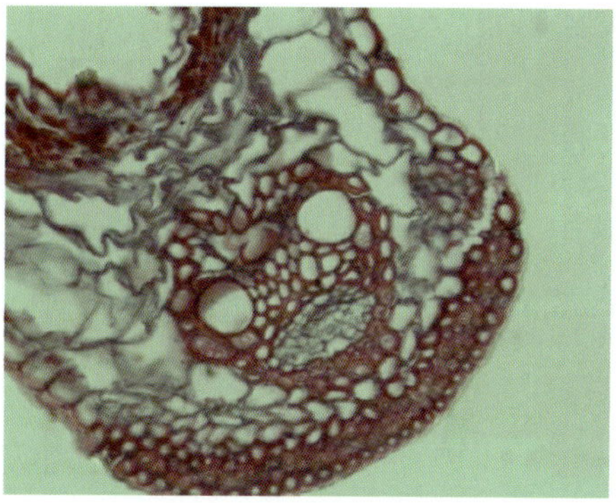

附图 7–2　淡竹叶横切片

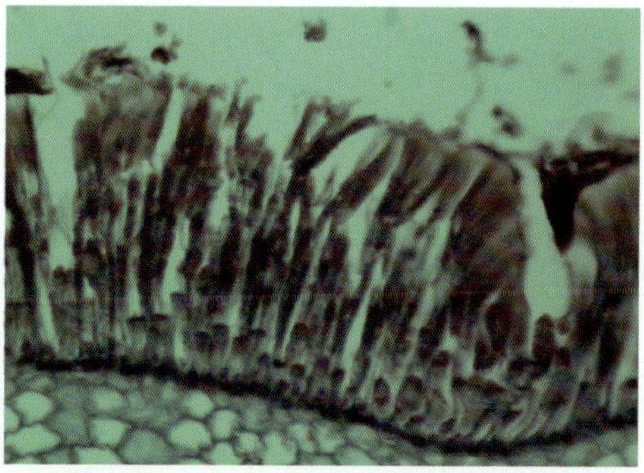

附图 10–1　海带带片横切示孢子囊

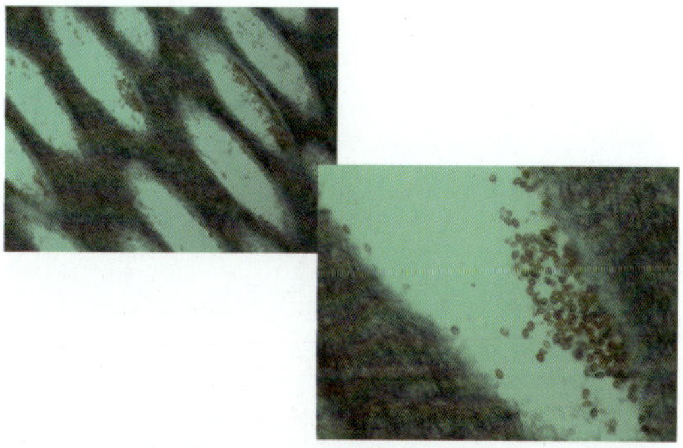

附图 10–2　灵芝菌盖内部构造

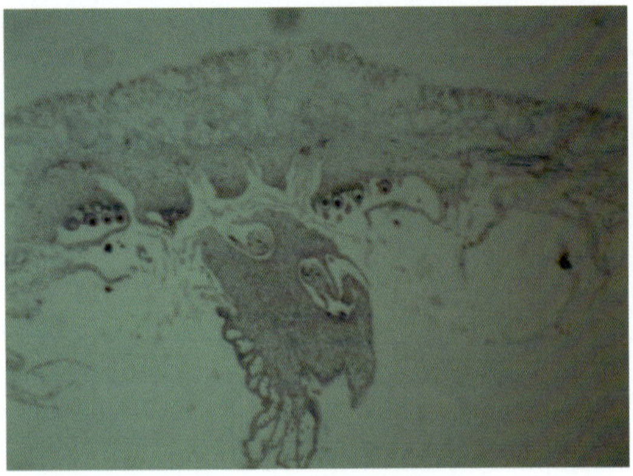

附图 11-1　地钱雌器托

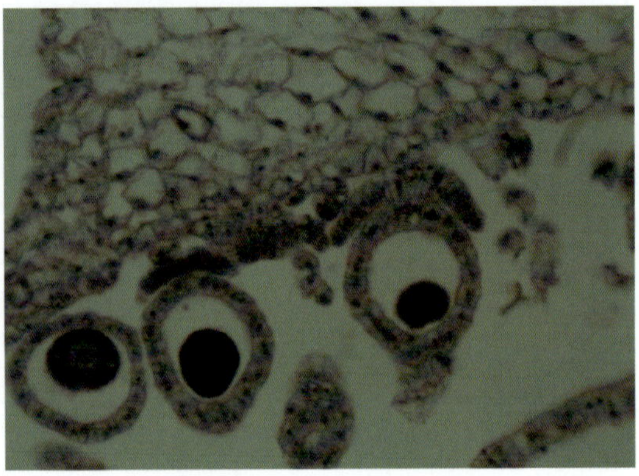

附图 11-2　地钱颈卵器

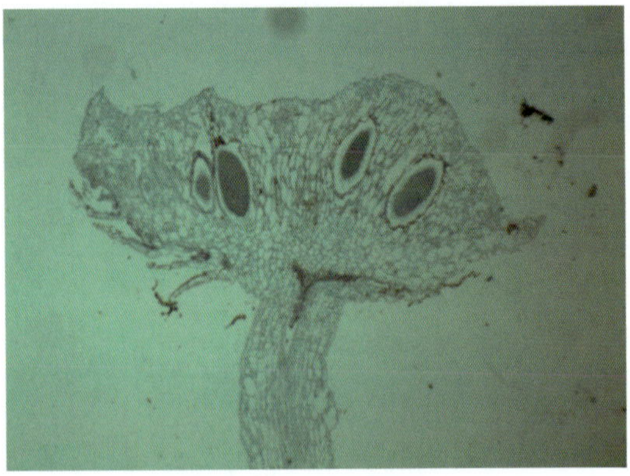

附图 11-3　地钱雄器托

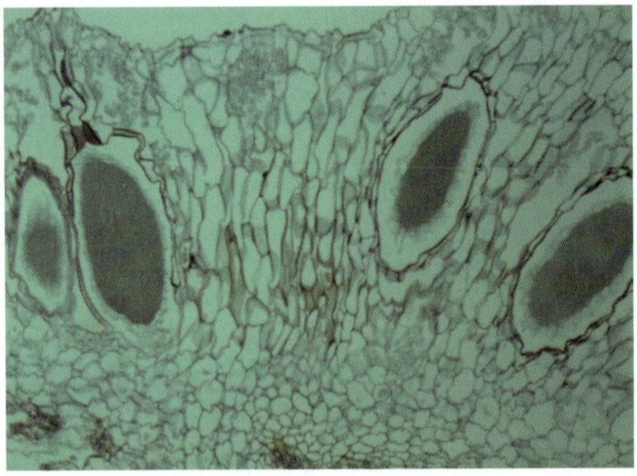

附图 11-4　地钱精子器

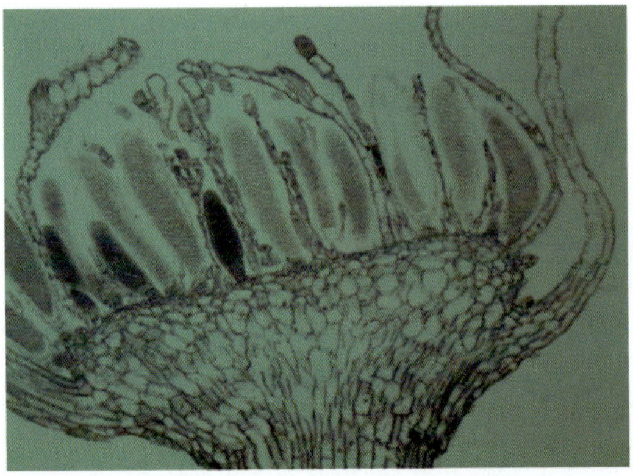

附图 11-5　藓的雄配子体

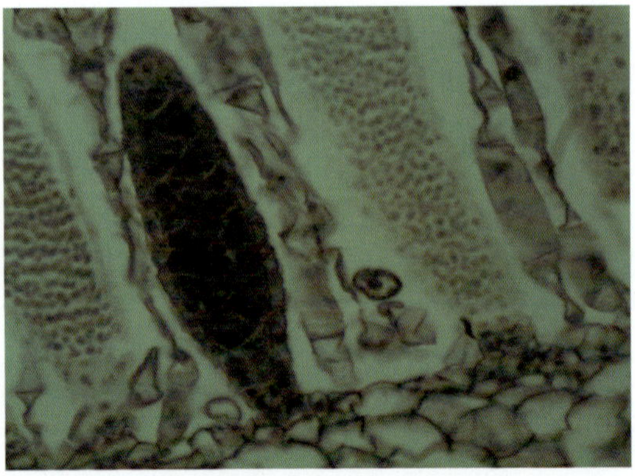

附图 11-6　藓的精子器

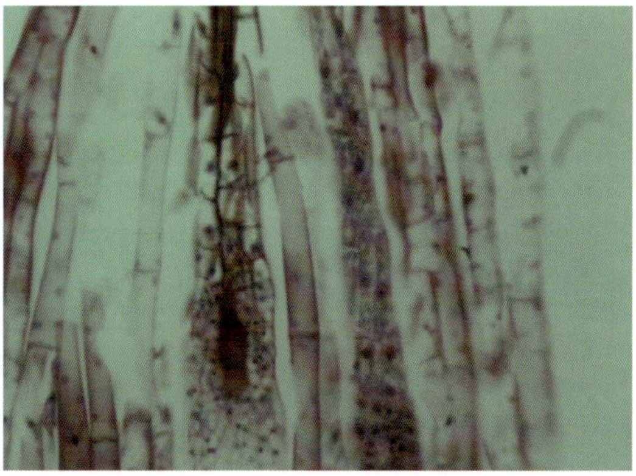

附图 11-7 藓的颈卵器

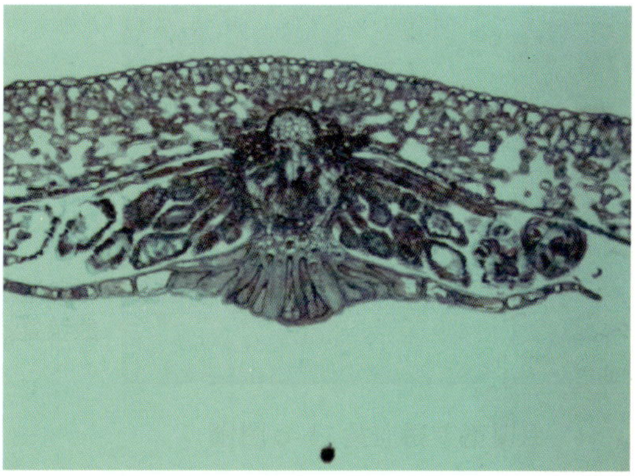

附图 11-8 粗茎鳞毛蕨孢子囊群

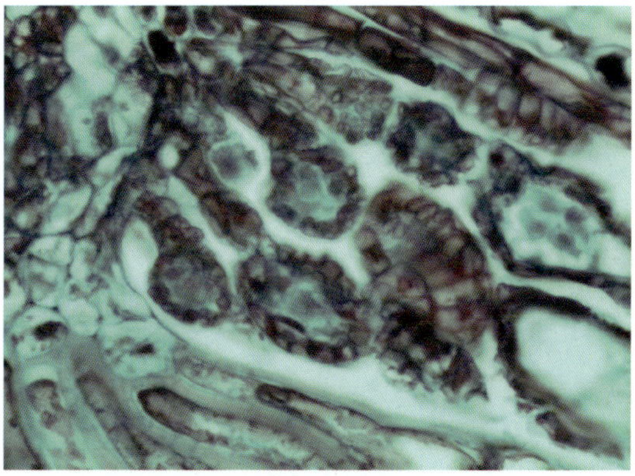

附图 11-9 粗茎鳞毛蕨孢子囊

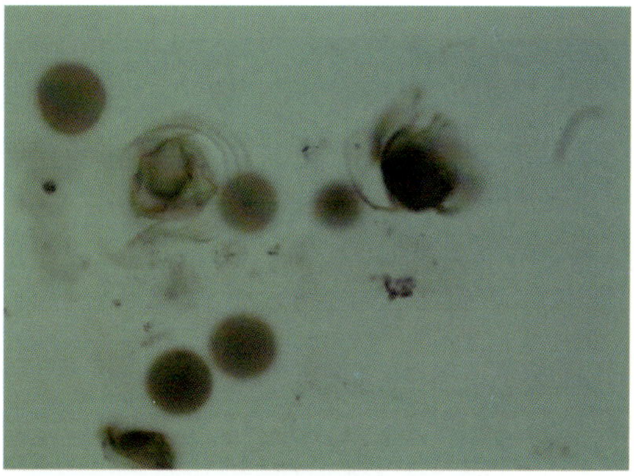

附图 11-10 木贼孢子示弹丝

附图 12-1 北细辛

附图 12-2 北乌头、花、果

附图 12-3 汉城细辛

附图 12-4 虎杖、花序

附图 13-1 五味子花、果

附图 13-2 芍药

附图 13-3 白屈菜

附图 13-4 牡丹

附图 13-5　延胡索

附图 13-6　菘蓝

附图 14-1　龙牙草

附图 14-2　山楂花、果

附图 14-3　山合欢

附图 14-4　甘草

附图 14-5　人参

附图 14-6　地榆

附图 14-7　膜荚黄芪

附图 14-8　苦参

附图 14-9　刺五加

附图 15-1　益母草

附图 15-2　曼陀罗

附图 15-3　柴胡

附图 15-4　丹参

附图 15-5　黄芩

附图 15-6　薄荷

附图 16-1　地黄

附图 16-2 桔梗

附图 16-3 红花

附图 16-4 蒲公英

附图 16-5 沙参

附图 16-6 党参

附图 16-7 北苍术

附图 16-7 紫菀

附图 17-1 薏苡

附图 17-2 半夏

附图 17-3 百合

附图 17-4 天南星

附图 17-5 麦冬

附图 18-1 薯蓣

附图 18-2 穿龙薯蓣